INJECTIVE MODULES AND INJECTIVE QUOTIENT RINGS

PURE AND APPLIED MATHEMATICS

A Program of Monographs, Textbooks, and Lecture Notes

Contributions to *Lecture Notes in Pure and Applied Mathematics* are reproduced by direct photography of the author's typewritten manuscript. Potential authors are advised to submit preliminary manuscripts for review purposes. After acceptance, the author is responsible for preparing the final manuscript in camera-ready form, suitable for direct reproduction. Marcel Dekker, Inc. will furnish instructions to authors and special typing paper. Sample pages are reviewed and returned with our suggestions to assure quality control and the most attractive rendering of your manuscript. The publisher will also be happy to supervise and assist in all stages of the preparation of your camera-ready manuscript.

LECTURE NOTES
IN PURE AND APPLIED MATHEMATICS

1. *N. Jacobson,* Exceptional Lie Algebras
2. *L.-Å. Lindahl and F. Poulsen,* Thin Sets in Harmonic Analysis
3. *I. Satake,* Classification Theory of Semi-Simple Algebraic Groups
4. *F. Hirzebruch, W. D. Newmann, and S. S. Koh,* Differentiable Manifolds and Quadratic Forms (out of print)
5. *I. Chavel,* Riemannian Symmetric Spaces of Rank One
6. *R. B. Burckel,* Characterization of C(X) Among Its Subalgebras
7. *B. R. McDonald, A. R. Magid, and K. C. Smith,* Ring Theory: Proceedings of the Oklahoma Conference
8. *Y.-T. Siu,* Techniques of Extension of Analytic Objects
9. *S. R. Caradus, W. E. Pfaffenberger, and B. Yood,* Calkin Algebras and Algebras of Operators on Banach Spaces
10. *E. O. Roxin, P.-T. Liu, and R. L. Sternberg,* Differential Games and Control Theory
11. *M. Orzech and C. Small,* The Brauer Group of Commutative Rings
12. *S. Thomeier,* Topology and Its Applications
13. *J. M. López and K. A. Ross,* Sidon Sets
14. *W. W. Comfort and S. Negrepontis,* Continuous Pseudometrics
15. *K. McKennon and J. M. Robertson,* Locally Convex Spaces
16. *M. Carmeli and S. Malin,* Representations of the Rotation and Lorentz Groups: An Introduction
17. *G. B. Seligman,* Rational Methods in Lie Algebras
18. *D. G. de Figueiredo,* Functional Analysis: Proceedings of the Brazilian Mathematical Society Symposium
19. *L. Cesari, R. Kannan, and J. D. Schuur,* Nonlinear Functional Analysis and Differential Equations: Proceedings of the Michigan State University Conference
20. *J. J. Schäffer,* Geometry of Spheres in Normed Spaces
21. *K. Yano and M. Kon,* Anti-Invariant Submanifolds
22. *W. V. Vasconcelos,* The Rings of Dimension Two
23. *R. E. Chandler,* Hausdorff Compactifications
24. *S. P. Franklin and B. V. S. Thomas,* Topology: Proceedings of the Memphis State University Conference
25. *S. K. Jain,* Ring Theory: Proceedings of the Ohio University Conference
26. *B. R. McDonald and R. A. Morris,* Ring Theory II: Proceedings of the Second Oklahoma Conference
27. *R. B. Mura and A. Rhemtulla,* Orderable Groups
28. *J. R. Graef,* Stability of Dynamical Systems: Theory and Applications
29. *H.-C. Wang,* Homogeneous Banach Algebras
30. *E. O. Roxin, P.-T. Liu, and R. L. Sternberg,* Differential Games and Control Theory II
31. *R. D. Porter,* Introduction to Fibre Bundles
32. *M. Altman,* Contractors and Contractor Directions Theory and Applications
33. *J. S. Golan,* Decomposition and Dimension in Module Categories
34. *G. Fairweather,* Finite Element Galerkin Methods for Differential Equations
35. *J. D. Sally,* Numbers of Generators of Ideals in Local Rings
36. *S. S. Miller,* Complex Analysis: Proceedings of the S.U.N.Y. Brockport Conference
37. *R. Gordon,* Representation Theory of Algebras: Proceedings of the Philadelphia Conference
38. *M. Goto and F. D. Grosshans,* Semisimple Lie Algebras
39. *A. I. Arruda, N. C. A. da Costa, and R. Chuaqui,* Mathematical Logic: Proceedings of the First Brazilian Conference
40. *F. Van Oystaeyen,* Ring Theory: Proceedings of the 1977 Antwerp Conference

41. *F. Van Oystaeyen and A. Verschoren,* Reflectors and Localization: Application to Sheaf Theory
42. *M. Satyanarayana,* Positively Ordered Semigroups
43. *D. L. Russell,* Mathematics of Finite-Dimensional Control Systems
44. *P.-T. Liu and E. Roxin,* Differential Games and Control Theory III: Proceedings of the Third Kingston Conference, Part A
45. *A. Geramita and J. Seberry,* Orthogonal Designs: Quadratic Forms and Hadamard Matrices
46. *J. Cigler, V. Losert, and P. Michor,* Banach Modules and Functors on Categories of Banach Spaces
47. *P.-T. Liu and J. G. Sutinen,* Control Theory in Mathematical Economics: Proceedings of the Third Kingston Conference, Part B
48. *C. Byrnes,* Partial Differential Equations and Geometry
49. *G. Klambauer,* Problems and Propositions in Analysis
50. *J. Knopfmacher,* Analytic Arithmetic of Algebraic Function Fields
51. *F. Van Oystaeyen,* Ring Theory: Proceedings of the 1978 Antwerp Conference
52. *B. Kedem,* Binary Time Series
53. *J. Barros-Neto and R. A. Artino,* Hypoelliptic Boundary-Value Problems
54. *R. L. Sternberg, A. J. Kalinowski, and J. S. Papadakis,* Nonlinear Partial Differential Equations in Engineering and Applied Science
55. *B. R. McDonald,* Ring Theory and Algebra III: Proceedings of The Third Oklahoma Conference
56. *J. S. Golan,* Structure Sheaves over a Noncommutative Ring
57. *T. V. Narayana, J. G. Williams, and R. M. Mathsen,* Combinatorics, Representation Theory and Statistical Methods in Groups: YOUNG DAY Proceedings
58. *T. A. Burton,* Modeling and Differential Equations in Biology
59. *K. H. Kim and F. W. Roush,* Introduction to Mathematical Consensus Theory
60. *J. Banas and K. Goebel,* Measures of Noncompactness in Banach Spaces
61. *O. A. Nielsen,* Direct Integral Theory
62. *J. E. Smith, G. O. Kenny, and R. N. Ball,* Ordered Groups: Proceedings of the Boise State Conference
63. *J. Cronin,* Mathematics of Cell Electrophysiology
64. *J. W. Brewer,* Power Series Over Commutative Rings
65. *P. K. Kamthan and M. Gupta,* Sequence Spaces and Series
66. *T. G. McLaughlin,* Regressive Sets and the Theory of Isols (in press)
67. *T. L. Herdman, S. M. Rankin, III, and H. W. Stech,* Integral and Functional Differential Equations
68. *R. Draper,* Commutative Algebra: Analytic Methods (in press)
69. *W. G. McKay and J. Patera,* Tables of Dimensions, Indices, and Branching Rules for Representations of Simple Lie Algebras
70. *R. L. Devaney and Z. H. Nitecki,* Classical Mechanics and Dynamical Systems
71. *J. Van Geel,* Places and Valuations in Noncommutative Ring Theory
72. *C. Faith,* Injective Modules and Injective Quotient Rings
73. *A. Fiacco,* Mathematical Programming with Data Perturbations I
74. *P. Schultz, C. Praeger, and R. Sullivan,* Algebraic Structures and Applications: Proceedings of the First Western Australian Conference on Algebra
75. *L. Bican, T. Kepka, and P. Nemec,* Rings, Modules, and Preradicals

Other Volumes in Preparation

INJECTIVE MODULES AND INJECTIVE QUOTIENT RINGS

Carl Faith

Rutgers, The State University
New Brunswick, New Jersey

MARCEL DEKKER, INC. NEW YORK and BASEL

Library of Congress Cataloging in Publication Data

Faith, Carl Clifton [date]
Injective modules and injective quotient rings.

(Lecture notes in pure and applied mathematics;
v. 72)
Includes bibliographical references and index.
1. Injective modules (Algebra) 2. Quotient rings.
I. Title. II. Series.
QA251.3.F34 512'.522 81-17495
ISBN 0-8247-1632-9 AACR2

MARCEL DEKKER, INC.
270 Madison Avenue, New York, New York 10016

Current printing (last digit):
10 9 8 7 6 5 4 3 2 1

PRINTED IN THE UNITED STATES OF AMERICA

PREFACE

These lectures are in two parts. Part I, entitled *Injective Modules Over Levitzki Rings*, studies an injective module E and chain conditions on the set $A_r(E,R)$ of right ideals annihilated by subsets of E. In analogy with the theorem of Hopkins and Levitzki, the Teply-Miller theorem states that the dcc implies the acc in $A_r(E,R)$. Moreover, a theorem of G. Hansen implies that E must have semiprimary biendomorphism ring Q (i.e., bicentralizer) in order for the dcc to hold in $A_r(E,R)$. In analogy with Σ-injectivity of E [which is equivalent to the acc in $A_r(E,R)$], we say that E is Δ-injective if the dcc holds in $A_r(E,R)$. Part I is devoted to proving the converse of the Teply-Miller-Hansen theorem: If E is Σ-injective and Q is semiprimary, then E is Δ-injective. Numerous applications are given in Chaps. 10 and 11, especially to commutative ring theory. In particular, in Chap. 11 theorems of I. Beck [33]' on when the injective hull E(R) of a commutative ring R is Σ-injective and the corresponding theorems for when E(R) is Δ-injective, are proved.*

Part I is as self-contained as practicable. It is a sequel to an earlier work,† and we refer to it for the double annihilator condition for injective and quasi-injective modules and the development of the maximal quotient ring; see Ref. 11', chap. 19, for a fuller treatment. We give a new proof, module-theoretically, of the Teply-Miller-Hansen theorem.

For a fuller discussion of Part I, see the Abstract and Introduction.

Part II is on the subject of (F)PF, or (finitely) pseudo-Frobenius, rings [i.e., all (finitely generated) faithful modules generate the category

---

*In the preface, I denote references listed in the first paper by the symbol [ ]' and references listed in the second paper by [ ]".

†C. Faith, *Lectures on Injective Modules and Injective Quotient Rings*, Lecture Notes in Mathematics 49, Springer-Verlag, Berlin-Heidelberg-New York, 1967.

mod-R of all R-modules]. (The PF rings had been introduced by Azumaya as a generalization of quasi-Frobenius rings, but FPF includes infinite products of Prufer domains, e.g., $\mathbb{Z}^{\omega}$.) In it, I verify all the conjectures that I made in Ref. 1" on commutative FPF rings.

The theorem is simply stated: A commutative ring R is FPF iff its classical quotient ring Q_c is injective, and every finitely generated faithful ideal of R is projective. The latter condition characterizes when R is (lower) pre-FPF; that is, every finitely generated faithful ideal of R generates mod-R. We say R is essentially upper right FPF if every finitely generated submodule of the injective hull E(R) of R that contains R generates mod-R. Then we prove the equivalence of the following:

FPF 1 R is FPF.

FPF 2 R is pre-FPF and essentially upper FPF.

FPF 3 R is quotient injective (i.e., Q_c is injective) and pre-FPF.

(I was well into the work before I saw that essentially upper FPF implies quotient-injective, and for that reason I did not see a reason to use the adjective "lower" on FPF. Actually, since I made no use of the concept, "upper pre-FPF," I could have used that terminology for "essentially upper FPF.")*

Part II is an application of techniques of noncommutative ring theory, and classical localization theory is required only in the sense of inverting regular elements. We do localize in the sense of Johnson-Utumi; that is, we use the maximal quotient ring $Q_m = Q_{max}(R)$. (When Q_c is injective, then $Q_m = Q_c$.) Because of the greater generality, Part II requires very little from Ref. 1" in the way of preparation. We do use the "rank-1" theorem from Ref. 1" stating that rank-1 generators are finitely generated projective. This means that any module M which generates mod-R and which embeds in Q_m has the stated property. This theorem is the key to our work, so we sketch another (simpler) proof in Part II.

A theorem of Azumaya states that any finitely generated faithful projective module P over a commutative ring R generates mod-R (the "rank-1" theorem is the converse for rank-1 modules). Thus, for example, R is essentially upper FPF if every finitely generated submodule of E(R) containing R is projective, and then as stated, R must be quotient-injective.

---

*In other words, FPF = upper and lower pre-FPF in this terminology.

A word about the organization of these notes. Each part is preceded by an abstract followed by a lengthier introduction which is largely independent of the abstract, but nevertheless, I recommend a reading of the abstract first. The organization of each paper is independent of the other, e.g., each supplied with its own bibliography, but we have supplied a common index.

C. Faith

CONTENTS

INJECTIVE MODULES AND INJECTIVE QUOTIENT RINGS

Part I

INJECTIVE MODULES OVER LEVITZKI RINGS

ABSTRACT An injective right R-module E is said to be Σ(resp. Δ)-*injective* if the set $A_r(E,R)$ of right ideals which are the annihilators of subsets of E satisfies the acc (dcc). The term Σ-injective derives from the characterization [1]: every direct sum $E^{(A)}$ of copies of E is injective, for any index set A. A theorem of Teply-Miller [2] states, in this terminology, that any Δ-injective module E is Σ-injective. Furthermore, then Q = Biend E_R, the biendomorphism or bicommutator ring of E, is semiprimary by a theorem of Hansen [4]. Let a *between ring for* E denote a subring S of Q containing the canonical image of $R \hookrightarrow Q$. We prove the converse of the Teply-Miller-Hansen theorem:

THEOREM An R-module E is Δ-injective iff E is Σ-injective and some between ring S for E is semiprimary. Then Q is semiprimary.

An injective R-module is *homogeneous* if it is a direct sum of copies of an indecomposable module. The next result also appears in a paper of Golan [9].

COROLLARY Then E is a finite direct sum of homogeneous injective modules.

By an example of Schelter-Small [3], Q need not be right artinian even if R is. The commutative ring case of the next result is due to Natasescu.

COROLLARY If some between ring S for E is right Kasch, e.g., if Q is primary-decomposable, then S = Q is right artinian.

The dual of the next corollary is known [1].

COROLLARY Every injective right R-module is Δ-injective iff R is right artinian.

Beck [33] proved that a commutative ring R has Σ-injective hull iff the classical quotient ring $Q_{c\ell}(R)$ is noetherian ([33], cor. 3.10).

COROLLARY R is noetherian (artinian) iff $Q_{c\ell}(R/A)$ is noetherian (artinian) for any ideal A.

The Cailleau-Beck characterization of Σ-injectives over commutative R together with the first two corollaries above yields:

COROLLARY Then an injective E over R is Δ-injective iff E is a direct sum of finitely many fully invariant homogeneous modules H_i, with H_i a direct sum of copies of the injective hull $E(R/P_i)$, where P_i is a prime ideal and R_{P_i} is an artinian local ring, i = 1, ..., n.

Examples: Any artinian (noetherian) injective module E over a commutative ring R is Σ(Δ)-injective.

When E is noetherian, then Q is artinian by the second corollary above, and the fact that E is finitely generated over Q implies that E defines a Morita duality context over Q.

Any direct summand of a direct sum or direct product of copies of a Σ(Δ)-injective module E is Σ(Δ)-injective.

1. INTRODUCTION

A right R-module M is an *upper* or *acc*-Levitzki module provided that R has acc on the set $A_r(M,R)$ of right ideals of R of the form

$$r_R X = \{r \in R \mid xr = 0, \forall x \in X\}$$

where $X \subseteq M$, $X \neq \emptyset$. By Ref. 1, M is acc-Levitzki iff for every right ideal I there corresponds a finitely generated right ideal I_1 with $\ell_M I_1 = \ell_M I$, where

$$\ell_M I = \{m \in M \mid mr = 0, \forall r \in I\}$$

An injective right R-module E is said to be Σ-*injective* provided any direct sum of copies of E is injective. It is known that E is Σ-injective iff E is upper Levitzki [1].

Dually, a module M is *lower* or *dcc-Levitzki* if $A_r(M,R)$ satisfies dcc, and M is *Levitzki* if both upper and lower. When R is a right upper (lower) Levitzki right R-module, we call R a right upper (lower) Levitzki ring. For uniformity of notation, we say that an injective module E is Δ-injective if E is lower Levitzki. (This is in purely formal analogy with Σ-injective, i.e., Δ is a formal symbol whose properties we propose to elucidate.) As stated, the theorem of Teply and Miller implies that Δ-injective $\Rightarrow$ Σ-injective, that is, lower implies upper Levitzki for injective modules [2].

Let Λ = End E_R denote the endomorphism ring of E, and let Biend E_R denote the biendomorphism (or double centralizer) ring, that is, Biend E_R = End ${}_\Lambda E$. In general, any quasi-injective R-module E is quasi-injective over Q = Biend E_R (Proposition 5.1D). When E is cofaithful over $\overline{R}$ in the sense that $\overline{R} = R/r_R E \hookrightarrow E^n$ for some $n < \infty$ (equivalently, $Q \hookrightarrow E^n$), then E is injective over any *between ring* S, that is, any ring S between $\overline{R}$ and Q (Corollary 5.6A). Furthermore, in this case, E is $\Sigma(\Delta)$-injective over R (i.e., Σ-injective, resp. Δ-injective) iff E is $\Sigma(\Delta)$-injective over S (Theorem 5.8B). (When E is Δ-injective, then E is cofaithful over $\overline{R}$.) Thus E(R) is $\Sigma(\Delta)$-injective over R iff it is over $Q \approx Q^r_{max}(R)$ (Theorem 5.7B).

It follows that a cofaithful injective E is $\Sigma(\Delta)$-injective whenever some between ring S is right noetherian (artinian). While it is known [3] that Q itself need not be artinian (noetherian), some sufficient conditions are that R be commutative, or that Q be right Kasch in the sense (9.1) that every maximal right ideal of Q is a right annihilator (Corollary 9.1A). The latter occurs when Q is a primary-decomposable semiprimary ring, e.g., when Q is a full $n \times n$ matrix ring over a completely primary ring.

In order for E to be Δ-injective, it is necessary for Q to be a semiprimary ring; that is, Q/rad Q must be semisimple artinian, and rad Q is nilpotent (Hansen's theorem [4]), and moreover, Λ = End E_R has nilpotent radical (Corollary 7.3). When E = E(R) is Δ-injective, then Λ itself is semiprimary [Theorem 8.4(5)]. [Compare the theorem of Harada and Ishii [5] which states that Λ is semiprimary when E is a quasi-injective module with $A_r(\Lambda,E)$ satisfying acc.]

By the double annihilator condition for quasi-injective modules [6] [stated in 5.3(1)], a (quasi-)injective module E is Δ-(quasi)-injective iff E as a module over Λ is noetherian, or as we say, iff E is counternoetherian. Thus the substance of the Teply-Miller theorem is the statement that any counternoetherian injective module is counterartinian, hence counter-Jordan-Hölder. In particular, then Q is the endomorphism ring of a module of finite length, a fact which generically implies semiprimary endomorphism ring (Theorem 6.2). [I do not know how Hansen proved that Q is semiprimary, but I ought to mention that the Harada-Ishii double annihilator condition, stated in 5.3(2), on the finitely generated left ideals of Λ, is used in proving *their* theorem.]

As stated, when some between ring for E is semiprimary, then E is Δ-injective iff E is Σ-injective over R (Theorem 8.9), and then Q is semiprimary. Actually the proof shows that a left perfect between ring S suffices, but then it is semiprimary; hence the well-known corollary: any right artinian ring R has semiprimary $Q^r_{max}(R)$. Another corollary is that every injective right R-module is Δ-injective iff R is right artinian. This is of course dual to the theorem for Σ-injective modules (see, e.g., Sec. 11.2), and indicates the "local" nature of the hypothesis on a single module. (See Theorem 8.9 ff., especially Corollary 8.13 and Theorem 8.15.)

A ring R is right Σ(Δ)-Levitzki, or right Σ(Δ) for short, if its injective hull E(R) in mod-R is Σ(Δ)-injective. Since every right ideal $I \in A_r(R)$ belongs to $A_r(M,R)$ for any faithful right module M, then a right Σ(Δ)-ring is upper (lower) right Levitzki, and hence a right Δ-ring R is a (right and left) Levitzki ring. Moreover, a theorem of Goursaud-Valette [7], reproved in the appendix to Chap. 5, asserts that any right Σ-ring is finite dimensional (i.e., satisfies acc ⊕ = the acc on direct sums of right ideals embeddable in R), and since R satisfies acc ⊥, then R is right Goldie.

A module H is *homogeneous* if H is a direct sum of isomorphic indecomposable modules. Of course, any completely decomposable module is a direct sum of homogeneous modules, but a Δ-injective module E is a direct sum,

$$E = \bigoplus_{i=1}^{n} H_i$$

of finitely many homogeneous completely decomposable injective modules $\{H_i\}_{i=0}^m$ (Theorem 8.6). These modules are determined as follows: There exist finitely many ideals $A_0, A_1, \ldots, A_n$ of $A_r(E,R)$ such that $E_{R/A_i}(R/A_i)$ (the injective hull of R/A_i in mod-R/A_i) is Δ-injective, i = 0, ..., n, and each indecomposable direct summand of E is isomorphic to a direct summand of some $E_{R/A_i}(R/A_i)$ (Theorem 8.6). It follows that $Q^r_{max}(R/A_i)$, $i = 0, \ldots, n$, and $Q = \text{Biend } E_R = \text{End } E$ are all semiprimary right Δ-rings, hence are right Goldie Levitzki rings.

By Theorem 5.8B, a module E is Δ-injective over R iff E is Δ-injective over $Q = \text{Biend } E_R$; hence a sufficient condition for an injective module E to be Δ-injective is for Q to be right artinian. Theorem 9.5 (10.8) shows that a right Δ-injective module E (Δ-ring R) has right artinian $Q = \text{Biend } E_R [Q = Q^r_{max}(R)]$ if any of the following hold:

($A\Delta_1$) R is commutative.

($A\Delta_2$) Q is right Kasch, i.e., every simple right ideal embeds in Q.

($A\Delta_3$) $J = J(Q) = \text{rad } Q$ belongs to $A_r(E,Q)$ [to $A_r(Q,Q)$].

($A\Delta_4$) $J^2 \in A_r(E,R)$ [$J^2 \in A_r(Q,Q)$].

($A\Delta_5$) The right socle $S_r = \text{soc}_r\, Q$ belongs to $A_r(E,Q)$ [$A_r(Q,Q)$].

($A\Delta_6$) S_r is an essential left ideal, equivalently, contains every minimal left ideal (hence contains $\text{soc}_\ell\, Q$).

($A\Delta_7$) Q is primary-decomposable (see below).

($A\Delta_8$) The prime ideals of Q commute.

By Asano's theorem a semiprimary ring Q is *primary-decomposable* i.e., decomposable into a product of finitely many primary rings (i.e., $n \times n$ full matrix rings over local semiprimary rings) iff the maximal ideals commute.

Moreover, since by Theorem 5.7B a ring R is a right $\Sigma(\Delta)$-ring iff $Q = Q^r_{max}(R)$ is a right $\Sigma(\Delta)$-ring, then e.g., Q right noetherian (artinian) suffices for R to be a right $\Sigma(\Delta)$-ring. Since Q is semiprimary, then in addition to $A\Delta_1$ through $A\Delta_8$, certain obvious conditions imposed on Q imply that Q is right artinian, e.g., the condition that Q is right noetherian, or (weaker) that $J(Q)/J(Q)^2$ has finite length [e.g., when $J(Q)^2 = 0$ (Corollary 9.6)].

Over a commutative ring R, then, E is Δ-injective iff $Q = \text{Biend }{}_R E$ is artinian; this happens, e.g., for any noetherian injective E [Theorems 9.5 (11) and 11.5.1]. Another necessary and sufficient condition for Δ-injectivity is $E = \oplus_{i=1}^n H_i$, where $n < \infty$ and H_i is a fully invariant homogeneous injective, i.e., H_i is of form $E_i^{(A_i)}$, where $E_i = E(R/P_i)$ is the injective hull of R modulo an "artinian" prime P_i (i.e., the local ring R_{P_i} is artinian). The proof uses Theorem 9.5 and the Cailleau-Beck theorem: if $E \approx \oplus_{i\in I} E_i$, where $E_i \approx E(R/P_i)$ for a prime ideal P_i such that R_{P_i} is a noetherian ring, and if E is injective, then E is Σ-injective, and conversely.

In Chap. 11, we give a partially new proof of this theorem which utilizes the fact that for any prime ideal P of a commutative ring R, E(R/P) is canonically an R_P-module which cogenerates the category mod-R_P of all R_P-modules. Then Σ-injectivity of E(R/P) over R implies that of E(R/P) over R_P, and since every ideal of R_P belongs to $A_r(E(R/P),R_P)$, the acc on the latter implies that R_P is noetherian. [Then, Matlis' theorem [8] implies that E(R/P) is an artinian R_P-module.] As in Beck's proof, in the case of a general Σ-injective we utilize Cailleau's theorem.

We say that R is a *fully right* $\Sigma(\Delta)$*-ring* if every factor ring is right $\Sigma(\Delta)$. A commutative ring R is fully $\Sigma(\Delta)$ iff R is noetherian (artinian) (Sec. 11.4). (Cf. Corollary 10.4B.)

In Chap. 7 a proof of the Teply-Miller theorem is given along module-theoretic and ring-theoretic lines by showing that an injective module E is lower Levitzki (i.e., Δ-injective) over R iff E has finite length over Λ = End E_R. The double annihilator condition for a (quasi-)injective module readily implies that E is lower Levitzki iff E is noetherian over Λ (i.e., counternoetherian), so the substance of this theorem is the proof that (1) E is counternoetherian iff (2) E is a counter-Jordan-Hölder module (i.e., has finite length over Λ). We prove this directly by showing:

(A) (1) implies that the Jacobson radical J = J(Λ) of Λ is nilpotent, so the chain $E \supset JE \supset \cdots \supset J^nE$ terminates in 0.

(B) There exists a finite integer n such that $\overline{E} = E(\overline{R})$ embeds in E^n, and since E^n is counternoetherian, then E^n, whence $\overline{E}$, is Δ-injective, where $\overline{R} = R/r_RE$. Thus, assume n = 1.

(C) The fact that $\overline{R}$ is essential in $\overline{E}$ implies that its endomorphism ring $\overline{\Lambda}$ is semiprimary inasmuch as $\overline{\Lambda}/\overline{J}$ is an epic image of $\overline{E}$.

(D) Therefore the noetherian modules $J^i\overline{E}/\overline{J}^{i+1}\overline{E}$ are semisimple, hence have finite length, and therefore so does $\overline{E}$.

(E) Write $\overline{E}$ = eE, for an idempotent $e \in \Lambda$. Then $\overline{\Lambda} \approx e\Lambda e$ is semiprimany, and therefore Λe is a Krull-Schmidt module in Λ-mod; in particular, $\Lambda e/Je$ is semisimple of finite length.

(F) Since $E = \Lambda\overline{R} \approx \Lambda/\Omega$, where $\Omega = \{t \in \Lambda \mid t(\overline{1}) = 0\}$, then $1. - e \in \Omega$; consequently $E \approx \Lambda e/\Omega e$ is an epic image of Λe.

(G) When J = 0, then E has finite length over Λ by (E) and (F). If J has index of nilpotency $n \neq 1$, and if the theorem is assumed for all injectives (over any ring) having the radical of its endomorphism ring nilpotent of index $< n$, then JE has finite length inasmuch as it is injective over R/r_RJE, and its endomorphism ring, being an epic image of Λ, has nilpotent radical of index $< n$.

After this was written, and presented at the Antwerp Ring Theory Conference 1978, I was acquainted with Hansen's theorem by J. Golan who refers to it in his own work [9]. Also, C. Nastasescu kindly made available a preprint [10] in which he places Theorem 7.3 in the setting of a Grothendieck category C with artinian generator, and gives a very simple proof. His theorem states that every artinian object of C is noetherian [[10], corollaire 1.2 (Hopkins)].

An earlier version of my proof of the Teply-Miller theorem rested on a false proposition stating that any noetherian module over a regular ring is artinian. This actually holds if R is commutative, or the module is finitely presented, but the proposition has been shown to be false by K. R. Goodearl, who has constructed left noetherian (artinian) modules over a regular ring $\bar{\Lambda}$ which are not artinian (noetherian). (But see Problems.)

2. ANNIHILATORS AND THE GALOIS CONNECTION

Let R and B be two rings, and M a (B,R)-bimodule. If X is a nonempty subset of either R or M, as stated, we let

$$r_R X = \{a \in R \mid xa = 0, \forall x \in X\}$$

If $X \subseteq R$, then $r_R X$ is an *annihilator right ideal*, or *right annulet* for short. If $X \subseteq M$, then $r_R X$ is a *right ideal*, and $A_r(M,R) = \mathrm{Ann}_r(M,R)$ will denote the *set of all* such right ideals. If M = R, then we set $A_r(R) = A_r(R,R)$. Dually, if $X \subseteq R$, then

$$\ell_M X = \{m \in M \mid mx = 0, \forall x \in X\}$$

is a B-submodule of M, and the set of these is denoted $A_\ell(M,R)$. If X is an ideal of R, then $\ell_M X$ is a (B,R)-submodule. Dually, if Y is a (B,R)-submodule of M, then $r_R Y$ is an ideal; in particular, then $r_R \ell_M X$ (resp. $\ell_M r_R Y$) is an ideal [(B,R)-submodule]. Now any R-module M is a (Λ,R)-bimodule, where $\Lambda = \mathrm{End}\, M_R$, and the (Λ,R)-submodules are called *fully invariant submodules*; hence $\ell_M X$ (resp. $r_R Y$) is a fully invariant submodule (resp. ideal) for any ideal X (resp. fully invariant submodule Y).

An ideal of $A_r(R)$ is called a *right annihilator ideal*, and an ideal of $A_\ell(R) = A_\ell(R,R)$ is called a *left annihilator ideal*.

If $Y \subseteq B$, then $r_M Y$ and

$$A_r(B,M) = \mathrm{Ann}_r(B,M) = \{r_M Y \mid Y \subseteq B\}$$

are defined by symmetry.

If A_ℓ and A_r are two lattices, and $g : A_\ell \longrightarrow A_r$, $h : A_r \longrightarrow A_\ell$ are order-reversing maps such that $hgX \supseteq X$ and $ghY \supseteq Y$ for all $X \in A_\ell$, $Y \in A_r$, then g induces a lattice anti-isomorphism $\bar{g}$ between the lattices $\bar{A}_\ell = \{hY \mid Y \in A_r\}$ and $\bar{A}_r = \{gX \mid X \in A_r\}$. Then $\bar{g}$ is called a *Galois connection* between A_r and A_ℓ (or $\bar{A}_r$ and $\bar{A}_\ell$) and $\bar{h} = \bar{g}^{-1}$. The objects of $\bar{A}_\ell$ are called the *Galois (closed) objects of* A, and similarly for $\bar{A}_r$. When A_ℓ and A_r are complete lattices, then so are $\bar{A}_\ell$ and $\bar{A}_r$ (see G. Birkhoff, *Lattice Theory*, Amer. Math. Soc. Colloquium Pub. Vol. 25, rev. ed., Providence, 1948). This fact is used in the proof of Prop. 4.4.

2.1 THE GALOIS CONNECTION FOR ANNIHILATORS If M is a right R-module, then annihilation defines a Galois connection between Λ-submodules of M and right ideals of R, which induces a lattice anti-isomorphism $A_r(M,R) \longrightarrow A_\ell(M,R)$ under which ideals maps onto fully invariant submodules. Moreover, $A_\ell(M,R)$ satisfies the ascending chain condition, denoted by acc $A_\ell(M,R)$, iff $A_r(M,R)$ satisfies the descending chain condition, denoted by dcc $A_r(M,R)$. Dually, $A_\ell(M,R)$ satisfies the descending chain condition, denoted dcc $A_\ell(M,R)$, iff $A_r(M,R)$ satisfies the ascending chain condition, denoted acc $A_r(M,R)$.

We let acc ⊥ R, or acc ⊥, represent acc $A_r(R)$, and then we say R has or satisfies acc ⊥. Dually, acc $A_\ell(R)$ is denoted by R ⊥ acc, or ⊥ acc, and we say R has ⊥ acc. Similarly for dcc ⊥ R and dcc ⊥, or R ⊥ dcc and ⊥ dcc. If $X \subseteq R$, then $r_R X$ is also denoted $X^\perp$, and $\ell_R X = {}^\perp X$.

2.2 PROPOSITION For any R-module M, ideal $A \subseteq r_R M$, and right ideal I containing A, it is true that $I/A \in A_r(M,R/A)$ iff $I \in A_r(M,R)$.

Proof: M is an R/A-module such that $x(r + A) = xr$, and the proposition is therefore obvious.

2.3 PROPOSITION (1) If E is a cogenerator of mod-R, then every right ideal of R belongs to $A_r(E,R)$.

(2) If E is injective, then the following are equivalent: (a) E cogenerates mod-R; (b) every simple R-module embeds in E; (c) every maximal right ideal belongs to $A_r(E,R)$; (d) every right ideal of R belongs to $A_r(E,R)$.

Proof: (1) See Ref. 11, proposition 23.13.

(2) (a) ⇔ (b), e.g., by Ref. 12, p. 148, corollary 3.31.

(b) ⇒ (c) If $h : R/I \hookrightarrow E$, then $I \in A_r(E,R)$, and in fact, $I = r_R x$, where $x = h(1 + I)$. Since every simple R-module $V \approx R/M$ for a maximal right ideal M, we have what we want.

(c) $\Rightarrow$ (b) Conversely, if $M = r_R X$ for a subset X of M, then maximality of M would imply that $M = r_R x$ for some $x \in X$, and hence $R/M \approx xR \subseteq E$ defines an embedding of the simple module R/M.

Now (a) $\Rightarrow$ (d) by (1), and (d) $\Rightarrow$ (c), so the proof is complete.

2.4 PROPOSITION If M is an R-module, then $I \in A_r(M,R)$ iff there is an embedding $h : R/I \hookrightarrow M^a$ for some product of copies of M. When this is so, then $I = r_R X$, where X is the set of all coordinates of $h(1 + I)$. Conversely, if $I = r_R X$, then $g : R/I \longrightarrow M^X$ is an embedding, where $g(r + I)(x) = xr$, for all $r \in R$, $x \in X$.

Proof: Trivial. Cf. Ref. 12, pp. 142-143.

3. LEVITZKI MODULES

A right R-module M is called *upper Levitzki*, or *acc-Levitzki*, when acc $A_r(M,R)$ holds. Dually for *lower* or *dcc-Levitzki*. M is *Levitzki* if both upper and lower Levitzki. If $B \longrightarrow \Lambda$ is a homomorphism of a ring B into $\Lambda = \text{End } M_R$, then M is canonically a (B,R)-bimodule, and by the Galois property, M will be upper Levitzki over R iff M has the dcc on B-submodules of $A_\ell(M,R)$. Then, in particular, M satisfies the dcc on the B-submodules in $A_\ell(M,R)$, for any subring B of End $_R M$. Conversely, if $A_\ell(M,R)$ satisfies the dcc on Λ-submodules, then M is upper Levitzki over R, and dually for lower Levitzki.

For this reason it is convenient to introduce the *countermodule* of M over R, namely, the module M considered as a module over its endomorphism ring. (The terminology is that of Stenstrom [13].) If the countermodule has property $\mathcal{P}$, then we say M is *counter-$\mathcal{P}$*, e.g., *counternoetherian*, *counterartinian*.

We say *finendo* in place of counter-finitely-generated. A *Jordan-Hölder* module is a module M of finite length $|M|$, that is, both artinian and noetherian.

3.1 PROPOSITION (1) An R-module M is upper (lower) Levitzki iff M is upper (lower) Levitzki over $R/r_R M$. In this case, any submodule N is upper (lower) Levitzki.

(2) If M is upper (lower) Levitzki over R, then for any ideal A or R, any submodule $N \subseteq \ell_M A$ is upper (lower) Levitzki over R/A.

(3) If M is counternoetherian(artinian), then M is lower (upper) Levitzki.

(4) If M is counter-Jordan-Hölder, then M is Levitzki.

Proof: (1) Let $A = r_R M$. By Proposition 2.2, for any right ideal I of R, it is true that $I \in A_r(M,R)$ iff $I \supseteq A$ and $I/A \in A_r(M,R/A)$, so $A_r(M,R)$ has acc (dcc) if $A_r(M,R/A)$ does. The last statement is obvious since $A_r(N,R) \subseteq A_r(M,R)$.

(2) By (1), N is upper (lower) Levitzki so we may assume M = N, and $A \subseteq r_R M$. Then (2) follows as in (1), employing Proposition 2.2.

(3) follows from the acc (dcc) on B-submodules of M, and the Galois correspondence; and (3) ⇒ (4).

3.2 PROPOSITION If M is an upper Levitzki faithful right R-module, then R is a right upper Levitzki ring, and satisfies acc ⊥. Dually, if R has a lower Levitzki faithful module, then R is right lower Levitzki, and satisfies dcc ⊥.

Proof: If I is a right annulet of R, then $I = r_R \ell_R I$. Since M is faithful, $r_R(M(\ell_R I)) = r_R \ell_R I = I$, and the proposition follows.

3.3 COROLLARY If M is upper (lower) Levitzki, then $R/\mathrm{Ann}_R M$ is a right upper (lower) Levitzki ring, hence satisfies acc ⊥ (dcc ⊥).

The *biendomorphism* ring of M, denoted Biend M_R, is End $_{\mathrm{End}\, M_R} M$, also called the *bicommutator* of R or *bicentralizer* of R in M. Thus, Biend M_R = End $_B M$, where B = End M_R.

3.4 PROPOSITION (1) A ring R is right upper (lower) Levitzki iff R embeds in a right upper (lower) Levitzki ring S.

(2) A right R-module M is upper (lower) Levitzki whenever M is upper (lower) Levitzki over any subring S of Biend M_R into which $R/r_R M$ embeds canonically. [Also, if $R \hookrightarrow S$, then $A_r(E,S)$ acc (dcc) implies $A_r(E,R)$ acc (dcc), for any S-module E.]

Proof: (1) If $I \subseteq E(R)$, then $r_R I = r_S I \cap R$, so acc ⊥ (dcc ⊥) in S implies that in R.

(2) Let $A = r_R M$. Since $R/A \hookrightarrow S$, then acc (dcc) in $A_r(M,S)$ implies acc (dcc) in $A_r(M,R/A)$. The proof of the parenthetical part is an exercise.

3.5 PROPOSITION For a right R-module M, the following are equivalent:

(1) M is upper (lower) Levitzki.

(2) Every direct sum of copies of M is upper (lower) Levitzki.

(3) Every direct product of copies of M is upper (lower) Levitzki.

Moreover, for any index set A, we have $A_r(M,R) = A_r(M^{(A)},R) = A_r(M^A,R)$, where $M^{(A)}$ is the direct sum.

Proof: Let X be a subset of the direct product M^A for an index set A. Then $r_R X = r_R(\cup_{a\in A} X_a)$ is the annihilator of the subset $\cup_{a\in A} X_a$, where X_a is the set of ath coordinates of X. Conversely, any $I \in A_r(M,R)$ belongs to $A_r(M^A,R)$, since M embeds in M^A. Since $M \subseteq M^{(A)} \subseteq M^A$, we get the desired equality for the A_r's.

See Theorem 5.19 for the applications to injective modules.

4. FINITE ANNIHILATORS

We next consider when $I \in A_r(M,R)$ has the form $r_R X$ for a finite subset of M; then I is called a *finite (right) annihilator.*

4.1 PROPOSITION A right ideal I of R is a finite right annihilator with respect to a right R-module M provided that the following conditions which are equivalent hold:

(1) $I = r_R X$ for a finite subset X of M.

(2) $I = r_R Y$ for a finitely generated Λ-submodule Y of M, where Λ = End M_R.

(3) $R/I \hookrightarrow M^n$ for some integer $n > 0$.

When this is so, then one may choose $|X| = n$.

Proof: (1) $\Rightarrow$ (2) using $Y = \Lambda X$, and (2) $\Rightarrow$ (1) using $X = \{x_1,\ldots,x_n\}$, a finite set of generators for Y over Λ. (1) $\Leftrightarrow$ (3) by Proposition 2.4.

A finite annihilator is *principal* in case n = 1 in the proposition.

4.2 COROLLARY (Dual to Proposition 4.1) A Λ-submodule N of M is a finite annihilator provided that the following equivalent conditions hold:

(1) $N = \ell_M I$ for a finite subset I.

(2) $N = \ell_M K$ for some right ideal K which is finitely generated modulo $r_R M$.

Proof: Same proof.

Caution: We use the term finite annihilator differently from Ref. 14, where a right R-submodule N of M is what we shall call *Beachy-Blair* finitely annihilated provided that $r_R N = r_R N'$ for a finite subset N' of N. In other words, in Beachy-Blair [14] the Galois correspondence $N \longmapsto r_R(N)$ is between fully invariant [i.e., (Λ,R)] submodules N and ideals of R.

4.3 PROPOSITION [1] (1) A right R-module M is upper Levitzki over a ring R iff for every right ideal I, there corresponds a finitely generated right ideal $I_1 \subseteq I$ with $\ell_M I = \ell_M I_1$.

(2) Dually, M is lower Levitzki iff for every Λ-submodule N there corresponds a finitely generated Λ-submodule N_1 with $r_R N = r_R N_1$.

This proposition follows from the next if we take I_1 in Proposition 4.3(1) to be the right ideal generated by the finite subset I' given below.

4.4 PROPOSITION If $h : A_\ell \longrightarrow A_r$, $g : A_r \longrightarrow A_\ell$ define a Galois connection between complete lattices A_ℓ and A_r, then A_r satisfies the acc on Galois objects iff for each $I \in A_r$ there corresponds a finite subset $I' \subseteq I$ such that $gI' = gI$.

Proof: Let I denote the l.u.b. of an ascending chain $I_1 \subseteq \cdots \subseteq I_n \subseteq \cdots$ in A_r, and let I' be a finite subset with $gI' = gI$. Then $I' \subseteq I_n$ for some n, and

$$hgI' = ghI = I \subseteq hgI_n = I_n \subseteq I$$

Conversely, if A_r has acc on Galois objects, A_ℓ has dcc on Galois objects, hence we may choose for any $I \in A_r$ a finite subset I' such that gI' is minimal in $\{gK \mid K \text{ finite subset of } I\}$. Then, since $g(I' \vee x) = gI'$, for all $x \in I$, $gI = \cap_{x\in I}\, g(I' \vee x) = gI'$ as claimed.

4.5 COROLLARY (1) If M is upper Levitzki over R, and if every finitely generated right ideal of R belongs to $A_r(M,R)$, then every right ideal belongs to $A_r(M,R)$ and R is right noetherian.

(2) Dually, if every finitely generated Λ-submodule of M belongs to $A_\ell(M,R)$, then M is lower Levitzki over R iff M is counternoetherian.

Proof: If I is any right ideal of R, then $\ell_M I = \ell_M I_1$ for finitely generated right ideal $I_1 \subseteq I$, and $\ell_R \ell_M I_1 = I_1 \supseteq I$, so $I = I_1$ is finitely generated, so R is right noetherian. Dually for (2).

Thus, any lower Levitzki quasi-injective module is counternoetherian (Proposition 5.4; cf. Theorem 7.1).

5. SIGMA QUASI-INJECTIVE MODULES

A *Baer homomorphism* of a right ideal I into M is a homomorphism $f : I \longrightarrow M$ "induced" by some element $m \in M$, that is,

$$f(x) = mx \quad \forall x \in M$$

Baer's criterion for injectivity of M states that M is injective iff every homomorphism of every right ideal of R into M is a Baer homomorphism (Baer [15]).

The *injective hull* E(M) of M is the *maximal essential extension* of M, or the *minimal injective overmodule* of M (Eckmann-Schopf [16]).

A module M is *quasi-injective* provided that every homomorphism $S \longrightarrow M$ of every submodule S of M into M is induced by an endomorphism of M. By a theorem of Johnson and Wong [6], M is quasi-injective iff M is a fully invariant submodule of E(M) of M, that is, $\lambda M \supseteq M$, for all $\lambda \in \text{End } E(M)_R$.

Note that ΛM is the least quasi-injective module H containing M, since $H = \Lambda H \supseteq \Lambda M$, and ΛM is quasi-injective by Johnson-Wong. The next result, due to Johnson-Wong [6] and Faith-Utumi [17], implies that every minimal quasi-injective over M is essential.

5.1A PROPOSITION Any module M has a unique (up to isomorphism) minimal quasi-injective essential extension QE(M), its quasi-injective hull. In fact, if $\Lambda = \text{End } E(M)_R$, then $QE(M) = \Lambda M$.

5.1B PROPOSITION (Harada [18]) If M is quasi-injective over R, so is M^n for any integer $n > 0$.

Proof: Write $E = E(M)$. Then $E^n = E(M^n)$, so $\text{End } E(M^n)_R = \Lambda_n$ for $\Lambda = \text{End } E_R$. Then $\Lambda_n M^n = M^n$, so M^n is quasi-injective.

For a complete study of when $M^{(\omega)}$ is quasi-injective, see Ref. 19. (Also, see the appendix at the end of this chapter.)

5.1C PROPOSITION If E is an injective module over R, then for any ideal A, $F = \ell_E A$ is a quasi-injective R-module and an injective canonical R/A-module.

Proof: We may assume $E(F) \subseteq E$. Then every $\lambda \in \text{End } E(F)_R$ is induced by $\lambda \in \text{End } E_R$. Since F is fully invariant in E, then it is in E(F), so F is quasi-injective. Consider $H = E_{R/A}(F)$. Since H is essential over F, we may assume $H \subseteq E$. But $HA = 0$ implies $H \subseteq F$, that is, $E_{R/A}(F) = F$ is injective in mod-R/A.

5.1D PROPOSITION If E is quasi-injective over R, then E is quasi-injective over $Q = \text{Biend } E_R$, in fact, over any subring S of Q containing the canonical image of $R/r_R E$ in Q.

Proof: If M is an S-submodule of E, then any morphism $f : M \longrightarrow E$ in mod-S induces a morphism f' of mod-R, hence extends to an endomorphism $\lambda \in \text{End } E_R = \text{End } E_S$ $(= \text{End } E_Q)$.

We let rad R denote the *(Jacobson) radical of a ring*, namely, the intersection of all maximal right ideals. See Ref. 11, chapter 18, and Ref. 20 for basic properties of rad R, including:

(J1) rad R is the intersection of all maximal left ideals.

(J2) $(1 + x)^{-1} \in R$ for all $x \in \text{rad } R$.

(J3) rad R contains every nil ideal of R.

(J4) $\text{rad}(R/\text{rad } R) = 0$.

A ring R is *(von Neumann) regular* provided that every principal (left) right ideal is generated by an idempotent; equivalently, for every $a \in R$ there corresponds $x \in R$ such that $axa = a$. (Note then $e = ax$ is an idempotent which generates aR.) Clearly, rad R = 0 in such a ring.

Let $K \subseteq' M$ indicate that K is an essential submodule of M. J(M) [resp. J(R)] will denote the radical of a module M (ring R).

5.2 THEOREM (Utumi [21]) If E is quasi-injective over R and $\Lambda = \text{End } E_R$, then $\Lambda/J(\Lambda)$ is a von Neumann regular ring, where

$$J(\Lambda) = \text{rad } \Lambda = \{\lambda \in \Lambda \mid \ker \lambda \subseteq' M\}$$

5.3 DOUBLE ANNIHILATOR CONDITION 1. (Jacobson [20], Johnson-Wong [6]) If M is a quasi-injective right R-module and if F is a finitely generated submodule of the countermodule, then F belongs to $A_\ell(M,R)$. Moreover, if a countersubmodule N of M belongs to $A_\ell(M,R)$, then so does N + F.

2. (Harada-Ishii [5]) Every finitely generated left ideal of B = End M_R belongs to $A_\ell(B,M)$.

5.4 COROLLARY A quasi-injective right R-module M is lower Levitzki iff M is counternoetherian.

Proof: One way is given by Proposition 3.1, and the converse follows from Proposition 4.4 and 5.3.

A module M is *cofaithful* provided that there is an integer $n > 0$ and an embedding $R \hookrightarrow M^n$. This is equivalent to the existence of elements $x_1, \ldots, x_n$ such that $\cap_{i=1}^n r_R x_i = 0$, and then $r \longmapsto (x_1 r, \ldots, x_n r)$ defines the required embedding. Any cofaithful module is faithful, and the next result states that any finendo faithful module module M is cofaithful. Over a right Artin ring R, every right module M is cofaithful over $R/r_R M$, and this property characterizes right Artin rings (see, e.g., Ref. 11, p. 69, theorem 19.16A, for references).

5.5 PROPOSITION If E is a finendo R-module, then E is cofaithful over $\overline{R} = R/r_R E$; in fact, if $\Lambda = \text{End } E_R$, then

$$\underset{\text{(in } \Lambda\text{-mod)}}{\Lambda^n \longrightarrow E \longrightarrow 0 \text{ exact}} \Rightarrow \underset{\text{(in mod-}\overline{R})}{0 \longrightarrow \overline{R} \longrightarrow E^n \text{ exact}}$$

In this case E^n is countercyclic. If E is quasi-injective over R, then conversely, E is cofaithful over $\overline{R}$ only if E is finendo.

Proof: Exactness of $\Lambda^n \longrightarrow E \longrightarrow 0$ converts via $\text{Hom}_\Lambda(\ ,E)$ into exactness of

$$0 \longrightarrow \text{Hom}_\Lambda(E,E) = Q \longrightarrow \text{Hom}_\Lambda(\Lambda^n,E) \approx E^n$$

If $A = r_R E$, then $R/A \hookrightarrow Q$ canonically, hence $R/A \hookrightarrow E^n$ as stated. If E is quasi-injective, then E is still quasi-injective over R/A (in fact, E is injective R/A; see Corollary 5.6A), and therefore $\text{Hom}_{R/A}(\ ,E)$ converts exactness of $0 \longrightarrow R/A \longrightarrow E^n$ into that of

$$\text{Hom}_{R/A}(E^n,E) = \text{Hom}_{R/A}(E,E)^n = \Lambda^n \longrightarrow E = \text{Hom}_{R/A}(R/A,E) \longrightarrow 0$$

It remains to prove that this implies that E^n is cyclic over $\Gamma = \text{End } E_R^n$. However, for *any* ring Λ and module E, exactness of $\Lambda^n \longrightarrow E \longrightarrow 0$ implies that of $\Lambda_n = \text{Hom}_\Lambda(\Lambda^n,\Lambda^n) \longrightarrow E^n = \text{Hom}\ (\Lambda^n,E) \longrightarrow 0$ via right exactness of $\text{Hom}_\Lambda(\Lambda^n,\ )$. Thus E^n is cyclic over $\Gamma \approx \Lambda_n$.

5.6A COROLLARY ([22] Let E be quasi-injective over R, and let $\overline{R} = R/r_R E \subseteq Q = \text{Biend } E_R$ canonically. If E is finendo, then E is injective over any subring S of Q containing $\overline{R}$.

Proof: Since Λ = End E_R = End E_S, then E is a finendo S-module, so by Proposition 5.5, $S \hookrightarrow E^n$ for some $n > 0$, and hence by Propositions 5.1B and D, any S-map $f : I \longrightarrow E^n$ of a right ideal I of S is induced by an element $t \in \Lambda$, and t induces an S-homomorphism $S \longrightarrow E^n$, which is Baer's criterion of injectivity of E^n over S. Thus, E is injective over S.

5.6B COROLLARY (1) Any finendo faithful quasi-injective module is injective.

(2) Any finendo quasi-injective module E is injective over Biend M_R.

Remark: Even if E is injective and simple, then E is not necessarily injective over its biendomorphism ring. (See the appendix of this section.)

M is a (Findlay-Lambek) *rational extension* of a submodule N provided that $\mathrm{Hom}_R(S/N,M) = 0$ for every submodule S of M containing N; equivalently, for each pair $x,y \in M$ and $y \neq 0$, there exists $r \in R$ with $xr \in N$ and $yr \neq 0$. (See, e.g., Ref. 23 or Ref. 11.)

5.7A THEOREM If E = E(R) and Λ = End E_R, then:

(1) (Findlay-Lambek) $\mathrm{Rat}^r_{\max} R = r_E \ell_\Lambda R$ is the maximal rational extension of R.

(2) (Lambek) Q = Biend E_R (= End ${}_\Lambda E$) is canonically isomorphic to $\mathrm{Rat}^r_{\max}(R)$ in mod-Q under $q \longrightarrow (1)q$, where 1 is the identity element.

(3) E = E(Q) in mod-Q.

Proof: Lambek [23], p. 94 ff., or Ref. 11, 19.35 ff.

We identify Q with $\mathrm{Rat}^r_{\max}(R)$, and designate this ring the *(Johnson-Utumi) maximal right* quotient ring $Q^r_{\max}(R)$. If R has a *classical right quotient ring* $Q_{c\ell}(R) = \{ab^{-1} \mid a,b \in R,\ b \text{ regular}\}$, then $Q^r_{\max}(R) \supseteq Q_{c\ell}(R)$ canonically [that is, $Q_{c\ell}(R) \subseteq \mathrm{Rat}^r_{\max} R$].

We recall the definitions given in the introduction: R is a right $\Sigma(\Delta)$-ring provided that E(R) is $\Sigma(\Delta)$-injective in mod-R.

5.7B THEOREM A ring R is a right Σ-ring (Δ-ring) iff $Q = Q^r_{\max}(R)$ is.

Proof: Suppose E = E(R) is Σ-injective in mod-R, and let $F = E_Q(E^{(a)})$. Then for $0 \neq y \in F$, we have $q \neq 0$ in Q with $0 \neq yq \in E^{(a)}$. Write $yq = (x_i) \in E^{(a)}$ with $x_i \in E$, and choose $r \in R$ and x_{i_0} such that $x_{i_0} r \neq 0$. Next choose $t \in R$ with $s = qrt \in R$ and $x_{i_0} rt \neq 0$. Thus $yR \cap E^{(a)} \neq 0$, so F is essential over $E^{(a)}$ in mod-R. Since $E^{(a)}$ is injective in mod-R, then $E^{(a)} =$ F is injective in mod-Q, so by (3) of Theorem 5.7A, Q is a Σ-ring.

Conversely, if Q is a Σ-ring (Δ-ring), then $A_r(E,Q)$ satisfies acc (dcc). Since every $I \in A_r(E,R)$ has the form $I_Q \cap R$ for $I_Q \in A_r(E,Q)$, then evidently $A_r(E,R)$ satisfies acc (dcc), proving that Q is a $\Sigma(\Delta)$-ring.

Next suppose R is a Δ-ring. Then, by Corollary 5.4, E = E(R) is counternoetherian. Since E = E(Q) in mod-Q and Λ = End E_Q, then in particular, E(Q) is counternoetherian; hence E = E(Q) is Δ-injective by Corollary 5.4; hence Q is a Δ-ring.

This theorem also follows from Theorem 5.8B.

(4) of the next theorem is essentially Lambek's observation in Ref. 24 that the biendomorphism ring B of any finendo faithful injective E ("nice" in his terminology) is a quotient ring of R, that is, $R \subseteq B \subseteq Q^r_{max}(R)$ canonically. Thus, if E is "small" (i.e., finitely generated over Λ), then B = Biend $_\Lambda E$ is "small", that is, embeds in Q.

5.7C THEOREM Let E be an injective R-module satisfying the equivalent conditions: (1) E is finendo; (2) E is cofaithful over $\overline{R} = R/r_R E$; (3) $\overline{R}$ embeds in E^n for some $n > 0$. Then

$$B \overset{(a)}{=} \text{Biend } E^n_R \overset{(b)}{=} \text{Biend } E_R \overset{(c)}{\hookrightarrow} \text{Biend } E(\overline{R})_R \overset{(d)}{=} Q^r_{max}(\overline{R}) \overset{(e)}{=} Q \qquad (4)$$

canonically.

Proof: Denote the respective relations by (a), (b), (c), (d), and (e) as indicated. Thus (a) is the definition of B, (b) holds by a classical matrix theorem (see, e.g., Proposition 5.15, (d) is essentially the definition of $Q = Q^r_{max}(\overline{R})$, using Biend $E(\overline{R})_R$ = Biend $E(\overline{R})_{\overline{R}}$. Thus it suffices to prove the theorem for the case $r_R E = 0$, that is, $R = \overline{R}$.

By Proposition 5.5, E^n is cyclic over Γ = End E^n_R. If $\gamma = \gamma^2 \in \Gamma$ is the projection idempotent for a decomposition $E^n = E(R) \oplus Y$ given by injectivity of E(R), say $\gamma | E(R) = 1_{E(R)}$, then

$$\gamma(xb) = (\gamma x)b = xb \qquad \forall b \in B,\ x \in E(R)$$

This shows that each b in B induces $\overline{b}$ in Q_2 = Biend $E(R)_R$, and $b \longmapsto \overline{b}$ defines a homomorphism $B \longrightarrow Q_2$ inasmuch as Λ = End $E(R)_R \approx \gamma\Gamma\gamma \subseteq \Gamma$, so that $\lambda(xb) = (\lambda x)b$ for all $\lambda \in \Lambda$ and all $b \in B$. Now $R \hookrightarrow B$ canonically, and since $R \hookrightarrow E(R)$, this implies that $B \hookrightarrow Q_2$ canonically [that is, xb = 0 for all $x \in E(R) \Rightarrow 1b = b = 0$].

Remark: Part of the proof just shows that any R-direct summand of E^n is a Q-direct summand. (Cf. also Proposition 5.15.)

Recall the definition of a *between ring* S for an injective module E given in the Abstract, namely, some ring such that $\overline{R} \subseteq S \subseteq Q$, where $\overline{R} = R/A$, $A = r_R E$, and $Q = \text{Biend } E_R$. This concept is used in Theorem 5.8B below.

An (quasi-)injective module E is Σ-(quasi-)*injective* iff every direct sum of copies of E is (quasi-)injective. For a proof of the next result, see Ref. 1 or Ref. 11, chapter 20. For (3) ⇔ (4), see Proposition 4.3.

5.8A THEOREM For an injective module E, the following are equivalent:

(1) E is Σ-injective.

(2) The (countable) direct sum $E^{(\omega)}$ is injective.

(3) E is upper Levitzki.

(4) Each right ideal I of R contains a finitely generated right ideal I_1 such that $\ell_E I_1 = \ell_E I$.

A quasi-injective module M is Σ-quasi-injective if E(M) is Σ-injective.

5.8B THEOREM Let E be injective over R, let $Q = \text{Biend } E_R$, and assume $\overline{R} = R/r_R E \subseteq Q$ canonically.

(1) E is Δ-injective over R iff E is Δ-injective over some between ring S. Then E is Δ-injective over every between ring for E.

(2) If E is Σ-injective over S, then E is Σ-injective over R, and conversely if E is finendo.

(3) A ring R is a right Σ-ring (Δ-ring) iff some between ring is a right Σ-ring (Δ-ring). In this case every subring of $Q \supseteq R$ is a right Σ-ring (Δ-ring).

Proof: By Corollary 5.4, E Δ-injective over R, or equivalently, by Proposition 3.1, Δ-injective over $\overline{R}$, implies that E is counternoetherian. Since E is thereby finendo, by Corollary 5.6A, E is injective over S in this case, hence Δ-injective over S, since E is a counternoetherian S-module. The converse is clear, e.g., apply Proposition 3.4(2).

(2) is proved similarly. If E is Σ-injective over S, then by Theorem 5.8A, E is upper Levitzki over S; hence E is upper Levitzki over R by Proposition 3.4(2), and then E is Σ-injective over R by Theorem 5.8A.

Conversely, if E is Σ-injective over R and finendo, and if $\Lambda = \text{End } E_R$, then by Proposition 5.5 there exists $n < \infty$ such that

$$\Lambda^n \longrightarrow E \longrightarrow 0 \quad \text{and} \quad 0 \longrightarrow \overline{R} \longrightarrow E^n$$

are exact, and then by Corollary 5.6A, E is injective over S. Let $F = E^{(\omega)}$

be the direct sum of ω copies of E. Since $\overline{R} \hookrightarrow F$ (via $E^n \hookrightarrow F$ and $\overline{R} \hookrightarrow E^n$), then F is (injective and) finendo. Since F is canonically an S-module which is faithful over S, and since End $F_R \approx \Lambda_{(\omega)}$, where $\Lambda_{(\omega)}$ is the ring of all row-finite matrices over Λ, then End F_R = End F_S and therefore $\overline{R} \subseteq S \subseteq$ Biend F_R canonically. Then Corollary 5.6A implies that F is injective over S.

(3) is immediate from Theorem 5.7B and the above.

For noetherian rings, the next theorem dates back to Matlis [8].

5.9 PROPOSITION (Cailleau [25]) For an injective module E the following are equivalent:

(1) E is Σ-injective.

(2) E is a direct sum of Σ-injective indecomposable modules.

The next result anticipates a statement of a result given in Chap. 6.

5.10 COROLLARY A ring R has a faithful finendo right Σ-injective module F iff E = E(R) is Σ-injective. In this case, $E = E_1 \oplus \cdots \oplus E_n$ is a direct sum of finitely many indecomposable injectives, and End E_R is a semiperfect ring.

Proof: ($\Rightarrow$) By Proposition 5.5, $R \hookrightarrow F^n$ for some $n > 0$, and hence F^n contains an injective hull E of R. Now F^n is Σ-injective (along with F); hence E(R) is Σ-injective, and by Proposition 5.2, E is a direct sum of indecomposable Σ-injective modules. Since R is finitely generated and essential in E, then $E = E_1 \oplus \cdots \oplus E_n$ is a finite direct sum of indecomposable Σ-injectives. Then by 6.1 (in the next chapter), End E_R is semiperfect. The converse follows from Proposition 5.5, since $R \hookrightarrow E(R) \Rightarrow E(R)$ is finendo.

5.11A PROPOSITION If E is Σ-injective (Δ-injective), then $F = \ell_E A$ is Σ-injective (Δ-injective) over R/A for any ideal A of R.

Proof: F is injective over R/A by Proposition 5.1C, and then F is Σ (resp. Δ)-injective by Proposition 3.1.

5.11B COROLLARY If E is a faithful Σ-injective R-module and A is an ideal, then $F = \ell_E A$ is a faithful Σ-injective module over R/A iff $A \in A_r(E,R)$.

Proof: Follows immediately from Proposition 5.11A since F is faithful over A iff $r_E F = A$, or, equivalently, iff $A \in A_r(E,F)$.

5.11C PROPOSITION If R is a right Σ-ring (resp. Δ-ring), then so is R/A for any ideal $A \in A_r(E(R),R)$.

Proof: If E = E(R) is Σ-injective, then E is finendo by Corollary 5.10. Since A is any ideal, $F = \ell_E A$ is fully invariant in E, and Σ-injective faithful over $\overline{R} = R/A$ by Corollary 5.11B, so by Theorem 5.7C, $E(\overline{R})$ embeds in E^n for an integer $n > 0$. Since E^n is Σ-injective, then by Proposition 3.1(2), $E(\overline{R})$ is Σ-injective over R, hence over $\overline{R}$, so $\overline{R}$ is a right Σ-ring.

If E = E(R) is Δ-injective, then it is finendo by Corollary 5.4, so by Theorem 5.7C again $E(\overline{R})$ embeds in E^n, and the rest of the proof is the same once we know that E^n is Δ-injective over R. However, E^n is counternoetherian, so Corollary 5.4 applies.

5.11D COROLLARY If E is injective over R and counternoetherian(-artinian), then E is Σ(Δ)-injective over R. If E is finendo, then E is Σ(Δ)-injective over any between ring S.

Proof: Apply Theorem 5.8A, Corollary 5.4, Proposition 3.1(3), and Theorem 5.8B.

5.11E REMARK The finendo condition is to assure that E is injective over S in the counterartinian case. Is this necessary? (Cf. Appendix to this chapter.)

5.11F REMARK By Proposition 3.1(3), when E is counternoetherian(-artinian), E will be lower (upper) Levitzki over S. In the former case E is finendo (by assumption), and in the latter case E is quasi-injective over S by Proposition 5.1D, so E will be Σ-injective over S whenever E is injective over S, e.g., when E is finendo (Corollary 5.6A).

A module M has *finite right Goldie dimension*, or G-dim $n < \infty$, iff M satisfies the following equivalent conditions:

GD1. M contains no infinite direct sum of nonzero submodules.

GD2. E(M) contains no infinite direct sums of nonzero submodules.

GD3. Every independent family of submodules is finite.

GD4. E(M) is a Krull-Schmidt module.

GD5. End $E(M)_R$ is a semiperfect ring (i.e., a semilocal lift/rad ring; see Chap. 6).

When this happens, then $E(M) = E_1 \oplus \cdots \oplus E_n$, where E_i is indecomposable, $i = 1, \ldots, n$, and n = G-dim M. We say then that M *satisfies* acc ⊕. (For proofs, see Ref. 12, p. 344, and Ref. 11, p. 45, proposition 18.26.)

For a proof of the next theorem, see the remark following Corollary 5.20.

5.12 PROPOSITION (Goursaud-Valette [7]) If R has a faithful Σ-injective right module, then R has finite right Goldie dimension.

A ring R is *right Goldie* provided that R satisfies acc ⊥ and acc ⊕, that is, R is right upper Levitzki of finite Goldie dimension.

5.13 COROLLARY A right Σ-ring R is right Goldie.

Proof: Apply Propositions 3.2 and 5.12.

Appendix

The question: When is a quasi-injective module E over R injective over $R/r_R E$? has been studied in various contexts. Fuller [26] gives one criterion which generalizes the finendo requirement (Proposition 5.5): The direct product E^A is quasi-injective for some set A of cardinal not less than the cardinal of a generating set of the countermodule (cf. Lambek [19], p. 179).

If all simple right R-modules are injective then R is said to be a right *V (for Villamajor) ring*. A V-ring without zero divisors is a *V-domain*.

5.14 EXAMPLE Let R be a right V-domain not a field [26a]. If E is any simple right module, then E is not injective over Q = Biend E_R. *Thus,* E *injective over* R *does not imply* E *injective over* Q.

Proof: The proof of this follows from the Chevalley-Jacobson density theorem and a result of Ref. 22. First, note that R ≠ Q since Q is a full linear ring, that is, the countermodule E is a vector space of infinite dimension, and E is a simple right Q-module, hence quasi-injective over Q. However, by Ref. 22, p. 163, proposition 19, E is not injective over Q. (Note that Q is left but not right self-injective, so even though E embeds in Q as a direct summand, E is not injective in mod-Q.)

In view of Proposition 5.7C, the following is interesting.

5.15 PROPOSITION If E is an R-module with $Q = \text{Biend } E_R$, then for any index set I, $Q \approx \text{Biend } E_R^{(I)}$ and $Q \approx \text{Biend } E_R^I$ canonically.

Proof: (Same proof as Ref. 12, p. 326, lemma 7.2, for $|I| < \infty$.) Let S denote the direct product or sum of $\{E_i\}_{i \in I}$, where $E_i = E$ for all i. Also, let e_i in End S_R induce the ith projection $S \longrightarrow E_i$. If $t \in P = \text{Biend } S_R$, then

$$xt = (e_i x)t = e_i(xt)$$

so $xt \in E_i$, and t induces an endomorphism $t_i \in \text{Biend } E_{i_R}$ (as was indicated in the proof of Proposition 5.7C); hence there exists $q_i \in Q$ such that $xt = xq_i$, for all $x \in E_i$. But q does not depend on i. For, if $u = e_{ji}$ is the automorphism of S which induces the canonical isomorphism $E_i \approx E_j$, and $y \in E_j = e_j E$, then $y = ux$ for $x \in E_i$, and hence $q_i = q_j$ for all $i,j \in I$, since

$$yq_j = yt = u(xt) = u(xq_i) = (ux)q_i = yq_i$$

This proves that the canonical map $Q \hookrightarrow P$ is an isomorphism, as stated.

5.16 PROPOSITION If E is $\Sigma(\Delta)$-injective over R, then E is Σ-quasi-injective (Δ-injective) over $Q = \text{Biend } E_R$.

Proof: If $F = E^{(I)}$ is injective over R, then by Proposition 5.1D, F is quasi-injective over Biend $F^{(I)}$, hence by Proposition 5.15, over Q. [Apply Corollary 5.6B(2) and Theorem 5.8B(1) for the Δ-case.]

5.17 EXAMPLE A direct sum of Δ-injective modules need not be Δ-injective (and hence neither is the product). Let R be a commutative von Neumann regular ring, e.g., the infinite product of fields, that is not noetherian. It is known that R_P is a field for all prime ideals P, so E(R/P) is always Δ-injective, but the direct sum E taken over all P is not even Σ-injective since then E would be a Σ-injective cogenerator of mod-R, in violation of a theorem [27] which would imply that R is noetherian. (See also Σ_5 in Sec. 11.2.)

Two right ideals A and B are *nested* if $A \supseteq B$. The next result, used in the proof of Theorem 8.9, generalizes lemma 3 of Ref. 1. (But, same proof!)

5.18 LEMMA If E is a Σ-injective module, then for each nested pair of right ideals A and B of R there exist finitely generated nested right ideals A_1 and B_1 such that $A \supseteq A_1$, $B \supseteq B_1$, $\ell_E A_1 = \ell_E A$, and $\ell_E B_1 = \ell_E B$.

Proof: By Theorem 5.8A, there exists finitely generated $A_1 \subseteq A$ such that $\ell_E A_1 = \ell_E A$. Since E is injective by proposition 23.21 of Ref. 11,

$$\ell_E(A_1 \cap B) = \ell_E A + \ell_E B = \ell_E B$$

so again by Theorem 5.8A, there exists finitely generated $B_1 \subseteq A_1 \cap B$ with $\ell_E B_1 = \ell_E B$.

5.19 THEOREM If E is $\Sigma(\Delta)$-injective, then so is any injective submodule F (equivalently, any direct summand) of any direct sum or product of copies of E.

Proof: One way is trivial, and the other way follows from Proposition 3.6 of the Δ-injective case, and the same result together with Theorem 5.8A in the Σ-injective case.

5.20 COROLLARY A ring R is a right $\Sigma(\Delta)$-ring iff there exists a faithful right $\Sigma(\Delta)$-injective module E.

Proof: R embeds in a product E^I of any faithful module, and E(R) embeds in such a product when E is injective, so the corollary follows from the theorem.

The corollary implies the Goursaud-Valette theorem, Proposition 5.12, since by Cailleau's theorem, Theorem 5.10, $E(R) = \oplus_{j\in J} E_j$ is a direct sum of indecomposable Σ-injective modules E_j $(j \in J)$, and since R is finitely generated (in fact cyclic), then $|J| < \infty$, which by the unique decomposition theorem, Theorem 6.1 of the next chapter, implies that R has finite Goldie dimension.

6. LEMMAS FROM FITTING-KRULL-SCHMIDT

A ring R is a *local ring* if it satisfies the equivalent conditions:

(L1) R/rad R is a (not necessarily commutative) field.

(L2) J = rad R is a maximal right ideal.

(L3) R has a unique maximal right ideal I.

(L4) The nonunits of R form an ideal N.

When this is so, then J = I = N. If E is an indecomposable right A-module, then End E_A is a local ring provided either that E is (1) quasi-injective, or (2) a Jordan-Hölder module over A.

(1) follows from the theorem of Utumi 5.2, and (2) is Fitting's lemma (N.B.).

A ring R is *semilocal* provided that R/J is semisimple artinian. A local ring R is semilocal, and so is the n × n matrix ring R_n over R.

A ring R is *semiperfect* provided that $R = \oplus_{i=1}^{n} e_iR$, where $e_i = e_i^2 \in R$, and e_iRe_i is a local ring, i = 1, ..., n. In this case, setting J(A) = rad A, we have $J(e_iRe_i) = e_iJe_i$, and $R/J \approx \oplus_{i=1}^{n} e_iR/e_iJ$ is a direct sum of simple modules e_iR/e_iJ, and hence then R is semilocal. One can characterize a semiperfect ring R as a *semilocal lift/rad* ring, where lift/rad means that each idempotent of R/J is the image under the canonical map R ⟶ R/J of an idempotent of R ([11], p. 42, proposition 18.23). If J is a nil ideal, then R is a lift/rad ring by a theorem of Jacobson [20] (e.g., Ref. 11, p. 41, 18.21).

The *basic ring* R_0 of a semilocal lift/rad ring R is defined by choosing an idempotent $e_0 = e_1 + \cdots + e_m$ in R such that, possibly after renumbering, every principal indecomposable right ideal e_jR is isomorphic to one and only one e_iR for $1 \leqslant i \leqslant m$. Then, tensoring over mod-R_0 by e_0R induces a category equivalence

$$\text{mod-}R_0 \longrightarrow \text{mod-}R$$

and R_0 is said to be *similar*, or *Morita equivalent*, to R ([11], proposition 18.24).

The *socle of a module* M, denote soc M, is the sum of all its simple submodules, or is 0. (It is a fully invariant submodule. If R is semilocal, then soc M = $\ell_M J$, where J = rad R.)

A module M is a *soc module* if soc M ≠ 0 or M = 0. A *fully soc module (Goldie module)* is one such that every factor module is a soc (Goldie) module. Clearly, every right R-module is a soc module iff every cyclic module is a soc module, or equivalently, R is a fully soc module in mod-R. In this case, we say R is a *right socular ring*, and then every right R-module M has an *ascending Loewy series* $\{S_\alpha\}_{\alpha\in\Gamma}$, where Γ is a well-ordered set, S_0 = soc M, $S_{\alpha+1}$ = soc M/S_α, and $S_\beta = \cup_{\alpha<\beta} S_\alpha$ for limit ordinals $\beta \in \Gamma$. Thus, either $M = S_{\alpha+1}$ for some α, or $M = S_\beta$ for a limit ordinal β. (The former obtains if M is finitely generated.) Dually, M has a descending Loewy series.

A ring R is *left perfect* if R has dcc on principal right ideals. Such a ring is characterized by Bass by the properties that R is (10 semi-perfect, and (2) right socular. (See Ref. 11, chapter 22.)

A ring R is *semiprimary* provided that R is semilocal and J is nil-potent. These rings are therefore right and left perfect.

A local semiprimary ring R is said to be *completely primary*, whereas a full n × n matrix R_n ring over R is said to be *primary*. Then a semiprimary ring R is *primary-decomposable* iff R is a finite product of primary rings. (See Theorem 9.8 for more on this.)

6.1 DEFINITION AND THEOREM A module M has an *Azumaya diagram* provided that (*) $M = \oplus_{i \in I} M_i$, where M_i is indecomposable and End $(M_i)_R$ is a local ring, for all $i \in I$. In this case the decomposition (*) is unique up to an automorphism of M (Azumaya [28]). If I is a finite index set, then (*) is called a *Krull-Schmidt diagram*, and M is called a *Krull-Schmidt module*. A module M is Krull-Schmidt iff End M_R is a semiperfect ring. (See, for instance, Ref. 11, p. 45, proposition 18.26.)

6.2 THEOREM If M is a module of finite length n over a ring R, then B = End M_R is a semiprimary Levitzki ring, and J = rad B is nilpotent of index $\leqslant n$.

Proof: First write $M = \oplus_{i=1}^n M_i$, where M_i is indecomposable, using the noetherian (or artinian) property. Then, by Fitting's lemma ([11], p. 17, proposition 17.17) M_i has local endomorphism ring, i = 1, ..., n, so the decomposition is a Krull-Schmidt diagram for M, and hence, as stated in 6.1, B is a semiperfect ring, and B is Levitzki by Proposition 3.2. Moreover, the same lemma (see, for instance, Ref. 11, p. 17, corollary 17.17') implies that any nil ideal of B is nilpotent of index $\leqslant n$, so it remains to prove that J is nil. Let $x \in J$. Then by Fitting's lemma again ([11], theorem 17.16), $M = M_1 \oplus M_2$, where x^t induces an automorphism in M_1, and $x^t M_2 = 0$ for some $t \leqslant n$. Write $M_1 = eM$ for $e = e^2 \in B$, and $M_2 = (1 - e)M$. Then $(exe)^t$ induces an automorphism of M_1, hence $(exe)^t$, whence exe, is a unit of eBe. But exe belongs to eJe = rad eBe, which implies eJe = eBe, so e is an idempotent in J, and hence e = 0. This implies that $x^t M = 0$, so x^t is nilpotent, and J is thereby a nil ideal, which by the Levitzki-Fitting theorem ([11], p. 19, theorem 17.20) implies nilpotency. Finally, B is Levitzki by the considerations beginning Chap. 3.

In general, when an R-module M has semiperfect, semiprimary, or semi-simple endomorphism ring B, then only limited information can be derived about the structure of M. For example, if R is a right Ore domain with right quotient field D, then D = End D_R canonically, but D is not finitely generated when R is not a field. Of course, in this case, D is a simple module over D = Biend D_R, as M is simple over Biend M_R whenever D = End M_R is a field. (This follows from the Chevalley-Jacobson density theorem.) However, when M is projective, one can say a bit more.

6.2A PROPOSITION If P is a projective left Λ-module which is a B-object in Λ-mod in the sense that every proper submodule is contained in a maximal subobject (e.g., when P is finitely generated), then for S = End $_\Lambda P$ and J = rad Λ we have rad S = $\text{Hom}_\Lambda(P,JP)$; S/rad S = End $_\Lambda P/JP$, and rad P = JP.

Proof: See, for example, Ref. 11, p. 154, especially 22.3 and 22.6; also p. 30, prop. 18.3.

A module M is a *local* module over Λ iff M has a unique maximal submodule; equivalently, iff M/rad M is simple. If $e = e^2 \in \Lambda$ and if $e\Lambda e$ is a local ring, then Λe is said to be a principal indecomposable left ideal, or *prindec* for short. Then a *principal cyclic module* is one which is a factor of a prindec. Every prindec, hence every principal cyclic module, is a local module, as the next result shows.

6.2B COROLLARY If P is a direct sum of left prindecs of Λ, then P/JP is semisimple.

Proof: If $P = \oplus_{i\in I} \Lambda e_i$ for prindecs Λe_i and $i \in I$, then P/JP = $\oplus_{i\in I} \Lambda e_i/Je_i$ is semisimple. (Note $e\Lambda e$ = End $_\Lambda \Lambda e$, and $e\Lambda e/eJe$ = End $_\Lambda \Lambda e/Je$ by Proposition 6.2A, for any idempotent $e \in \Lambda$. Now $\Lambda e/Je$ embeds as a left ideal in the ring Λ/J under the canonical map, and it is known [20] that a left ideal of a semiprime ring is simple iff its endomorphism ring is a field. Since $e_i\Lambda e_i/e_iJe_i$ is a field, then $\Lambda e_i/Je_i$ is simple, for every $i \in I$.)

6.3 PROPOSITION If E is a finitely generated Σ-injective module over R, then B = End E_R is a right perfect ring, and E is cofaithful over B.

Proof: By the proof of Theorem 5.7C, there is an embedding $B \hookrightarrow E^n$ in B-mod, where $R^n \longrightarrow E \longrightarrow 0$ is exact in mod-R. By the double annihilator condition 5.1A, E satisfies [via Theorem 5.8A(3) and Proposition 3.1] the dcc on finitely generated B-submodules. Since the latter property is preserved by direct sums [29], then E^n, whence B, satisfies the stated dcc.

6.4 COROLLARY A finitely generated Δ-injective module E has left artinian endomorphism ring B.

Proof: By Theorem 7.1, E has finite length over B, so B has finite length in B-mod by Proposition 6.3.

6.5 PROPOSITION If E is an injective R-module which is noetherian (artinian) over Q = Biend E_R, e.g., if E is noetherian (artinian), then B = End E_R is semiprimary (left noetherian and semiperfect).

Proof: By the theorem of Harada and Ishii, 5.3(2), the stated condition on E implies the dcc (acc) on finitely generated left ideals of B, and therefore B is right perfect (left noetherian). (In either case, B is semiperfect by 6.1.) Now $r_E J^n = r_E J^{n+1}$ for some finite n, but if Y denotes this submodule, and if S/Y is the socle of E/Y in B-mod, then $JS \subseteq Y$ implies that $J^n(JS) = 0$, so then $J^n S = 0$, where J = rad B. Then $S \subseteq Y$, and the fact that B is left socular implies that E = Y and that $J^n = 0$, proving that B is semiprimary.

6.6 COROLLARY Let R be a semiprimary ring with radical J, and let n be the index of nilpotency of J. Then the following are equivalent:

(1) R is right artinian.

(2) R is right noetherian.

(3) J^i/J^{i+1} has finite length (equivalently, finite Goldie dimension), i = 1, ..., n, as a right R-module.

(4) R/J^i has finite right Goldie dimension, i = 1, ..., n.

Proof: Exercise.

7. THE TEPLY-MILLER THEOREM

As stated in the introduction, the Teply-Miller theorem is a consequence of the next theorem. The proof requires very little of the formalism of the preceding sections, most of which was developed for succeeding sections.

It is shown that the length of any maximal chain of right ideals in $A_r(E,R)$ is an invariant of E, called the *Levitzki length* of E (see Corollary 7.4).

7.1 THEOREM A counternoetherian quasi-injective (i.e., Δ-quasi-injective) right R-module E is counterartinian (i.e., Σ-quasi-injective).

Proof: Let $A = r_R E$. Then by Proposition 5.5 there exists $n > 0$ such that $R \hookrightarrow E^n$, in which case E^n is cyclic over its endomorphism ring $\Gamma = \Lambda_n$, where $\Lambda = \text{End } E_R$. Since $\text{Hom}_\Lambda(\Lambda^n, \ )$ induces a category equivalence,

$$\text{Hom}_\Lambda(\Lambda^n, \ ) : \Lambda\text{-mod} \approx \Lambda_n\text{-mod}$$

we see that E is noetherian (artinian) in Λ-mod iff E^n is noetherian (artinian) in Λ_n-mod. Thus, in order to prove the theorem, it suffices to prove it for the case where E is cyclic over Λ, or equivalently, when R/A embeds in E. Also, since E is also quasi-injective (in fact, injective) over R/A, we may as well assume A = 0, that is, prove the theorem for E faithful. (Actually, this is just a notational convenience to allow us to write R in place of R/A.)

By Proposition 5.5, $E = \Lambda 1 = \{\lambda(1) \mid \lambda \in \Lambda\}$; that is, the cyclic Λ-module E is generated by the unit element $1 \in R$. Let $\Omega = \{\omega \in \Lambda \mid \omega 1 = 0\}$, making

$$0 \longrightarrow \Omega \longrightarrow \Lambda \longrightarrow E \longrightarrow 0 \tag{7.1}$$

exact in Λ-mod. If Δ is any ideal of Λ, then by the noetherian hypothesis on $E \approx \Lambda/\Omega$, there exist finitely many elements $\lambda_1, \ldots, \lambda_n \in \Delta$ which generate Δ modulo Ω, that is,

$$\Delta = \sum_{i=1}^{n} \Lambda\lambda_i + \Omega \tag{7.2}$$

Since

$$\Delta E = \Delta\Lambda 1 = \Delta 1 = \{\delta(1) \mid \delta \in \Delta\} \tag{7.3}$$

and since $\omega(1) = 0$, for all $\omega \in \Omega$, we have

$$\Delta E = \sum_{i=1}^{n} \Lambda\lambda_i(1) \tag{7.4}$$

Therefore,

$$r_R(\Delta E) = \bigcap_{i=1}^{n} \ker(\lambda_i \cap R) = r_E\Delta \cap R \tag{7.5}$$

Now the Double Annihilator Theorem 5.3 implies that

$$\Delta E = \ell_E r_R(\Delta E) = \ell_E(r_E\Delta \cap R) \tag{7.6}$$

And consequently, we can conclude, for any two ideals Δ_1 and Δ_2 of Λ, that

$$r_E\Delta_1 = r_E\Delta_2 \Leftrightarrow \Delta_1 E = \Delta_2 E \tag{7.7}$$

Since E is noetherian in Λ-mod, the chain

$$r_E J \subseteq \cdots \subseteq r_E J^n \subseteq \cdots \tag{7.8}$$

terminates; hence by (7.7) there exists some integer n such that

$$J^n E = J^{n+1} E = J(J^n E) \tag{7.9}$$

and then Nakayama's lemma implies $J^n E = 0$, and hence $J^n = 0$. Thus,

$$E \supseteq JE \supseteq \cdots \supseteq J^{n-1}E \supseteq 0$$

is a chain of fully invariant submodules of E. We propose to prove: (1) $|E/JE| < \infty$; and (2) $|JE| < \infty$, as left Λ-modules. We first establish:

7.1A LEMMA If $\overline{E} = E(\overline{R})$ is Δ-injective, then $\overline{\Lambda} = \operatorname{End} \overline{E}_R$ is semiprimary.

Proof: To simplify notation, omit the bars. Since R is essential in E, then Ω is contained in J, and hence Λ/J is an epic image of $E \approx \Lambda/\Omega$. Since Λ/J is thereby noetherian, and von Neumann regularity implies that Λ/J is semisimple artinian, and since J is nilpotent, then Λ is semiprimary as stated.

The lemma applies to the proof of the theorem since $E(\overline{R}) \subseteq E$ is Δ-injective.

To complete the proof of the theorem, we choose an idempotent $e \in \Lambda$ which induces the projection of E onto $\overline{E}$, in a direct sum decomposition $E = \overline{E} \oplus X$. Thus, $(1 - e)\overline{E} = 0$, and therefore $1 - e \in \Omega$; it follows from this that $E \approx \Lambda/\Omega \approx \Lambda e/\Omega e$, so E is an epic image of Λe in Λ-mod. Since $e\Lambda e \approx \operatorname{End} eE_R$ (= End $\overline{E}_R$ canonically) is semiprimary, hence semiperfect, then $e = e_1 + \cdots + e_n$ for mutually orthogonal idempotents e_i of $e\Lambda e$ such that $e_i \Lambda e_i$ is a local (and semiprimary) ring. Then Λe is a direct sum of left prindecs, $\Lambda e = \oplus_{i=1}^n \Lambda e$, and by Corollary 6.2B, $\Lambda e/Je$ is a semisimple module. Now $E/JE = \Lambda e/(Je + \Omega e)$ is an epic image of $\Lambda e/Je$, hence semisimple. This proves the theorem for the case $J = 0$, and also statement (1).

We complete the proof by induction on the index n of nilpotency of the endomorphism ring of a Δ-injective module. Since JE is fully invariant in E, one sees that JE is quasi-injective over R, and that its endomorphism ring Λ_1 is an epic image of Λ; moreover, J maps onto the radical of Λ_1 since by Utumi's theorem, rad Λ_1 consists of endomorphisms with essential kernels. Moreover, by Corollary 5.6A, JE is an injective module over $R_1 = R/r_R JE$, and hence JE is a Δ-injective module (e.g., since it is counternoetherian), with the index of nilpotency of the radical J_1 of its endomorphism ring Λ_1 not exceeding n - 1. By the induction hypothesis, then (2) holds, and hence $|{}_\Lambda E| < \infty$.

7.2 COROLLARY (Teply-Miller [2]) Any lower Levitzki injective module E is Levitzki, that is, dcc ⇒ acc in $A_r(E,R)$.

As stated, part of the next result is in Hansen's thesis.

7.3 COROLLARY If E is Δ-injective over R, then Λ = End E_R has nilpotent radical, and $Q = Q^r_{max}(\overline{R})$ = End ${}_\Lambda E$ is semiprimary, where $\overline{R} = R/r_R E$.

Proof: Since E has finite length n over Λ, then obviously $J(\Lambda)^n = 0$. Moreover, Q is semiprimary by Theorem 6.2.

The Levitzki Length of a Module

Let E be injective over R and Λ = End E_R. A chain $\{I_i\}$ of right ideals (Λ-submodules) of R (of E) is a *Levitzki chain* if $I_i \in A_r(E,R)$, for all i $[I_i \in A_\ell(E,R)]$.

7.4 COROLLARY If E is injective over R, then all Levitzki chains of $A_r(E,R)$ are finite iff E is Δ-injective.

Then the length n of the countermodule of E is the length of any maximal Levitzki chain, and is called the *Levitzki length of* $A_r(E,R)$.

Proof: Clear from the proof of Theorem 7.1, the Galois connection $A_\ell(E,R) \longrightarrow A_r(E,R)$, and the fact that $A_\ell(E,R)$ contains every Λ-submodule of E.

7.5 COROLLARY An injective module E is Δ-injective iff E is a counter-Jordan-Hölder module.

Proof: By Corollary 5.4, E is Δ-injective iff counternoetherian, and hence, by Theorem 7.1, iff E is a countermodule of finite length.

8. Δ-INJECTIVE MODULES

In this chapter the structure of a Δ-injective module E is determined in Theorem 8.6, as outlined in the Introduction, e.g., there are only finitely many homogeneous direct summands, and as a corollary, only finitely many fully invariant direct summands. Also, as stated in the Abstract, a module E is Δ-injective iff E is Σ-injective with a semiprimary between ring (Theorem 8.9). In fact, a left perfect (i.e., right Björkian) between ring will do (Corollary 8.12). Finally, every right R-module is Δ-injective iff R is right artinian (Theorem 8.16).

8.1 PROPOSITION Let E be a quasi-injective module. The following are equivalent:

(1) E is Δ-quasi-injective.

(2) E is counternoetherian.

(3) E is counter-Jordan-Hölder.

Proof: This is a restatement of the results of Chap. 7.

8.2 COROLLARY If E is Δ-quasi-injective, so is E^n, for any $n > 0$. (Compare Theorem 5.18.)

Proof: E^n is quasi-injective by Proposition 5.3(2). Moreover, as in the proof of Theorem 7.1, E^n is noetherian over Λ_n = End E^n_R.

8.3 COROLLARY Let a quasi-injective R-module E satisfy the three equivalent conditions:

(A) E is Δ-quasi-injective over R.

(B) E is noetherian over Λ = End E_R.

(C) E has finite length over Λ.

Then:

(1) J(Λ) = rad Λ is nilpotent of index $\leqslant n = |_{\Lambda}E|$.

(2) Q = End $_{\Lambda}E$ is a semiprimary Levitzki ring with nilpotent radical of index $\leqslant$ n.

Proof: (1) This holds for any faithful module E of finite length (as in the proof of Theorem 7.1). (2) Apply Theorem 6.2.

8.4 THEOREM Let E be Δ-quasi-injective over R, and let $\overline{R} = R/r_R E$.

(1) There exists an integer $n > 0$ such that $E^n = QE(R) \oplus X$ (in mod-R).

(2) $\overline{R}$ has semiprimary Levitzki maximal right quotient ring Q = $Q^r_{max}(\overline{R})$.

(3) The quasi-injective hull QE($\overline{R}$) of $\overline{R}$ in mod-R is Δ-injective both as $\overline{R}$- and $\overline{Q}$-modules.

(4) QE($\overline{R}$) is the injective hull of $\overline{R}$ in mod-R, and of $\overline{Q}$ in mod-$\overline{Q}$.

(5) Λ = End $E(\overline{R})_R$ is a semiprimary ring.

Proof: (1) By Proposition 5.5, $\overline{R}$ embeds in E^n for an integer $n > 0$, and E^n is also counternoetherian. Moreover, an injective hull $E(\overline{R})$ of the $\overline{R}$-module $\overline{R}$ is contained in E^n; in fact, is a direct summand.

Write $E(\overline{R}) = eE^n$ for an idempotent $e \in \Gamma = \text{End } E^n_R$. Then $\Lambda = e\Gamma e \approx \text{End } E(\overline{R})_R$ canonically. If M is any Λ-submodule of $E(\overline{R})$, then $\Gamma M = \Sigma_{i=1}^m \Gamma x_i$ is finitely generated by, say, $x_1, \ldots, x_m$ in M, and hence

$$e\Gamma M = \sum_{i=1}^{m} e\Gamma em_i = \sum_{i=1}^{m} \Lambda m_i$$

is finitely generated over Λ; that is, $E(\overline{R})$ also is counternoetherian, hence counterartinian by Theorem 7.1, hence counter-Jordan-Hölder. Now $QE(\overline{R})$ is contained in $E(\overline{R})$, and is trivially a quasi-injective $\overline{R}_i$ module which must be injective in mod-$\overline{R}$ by Corollary 5.6A; that is, $QE(\overline{R}) = E_R(\overline{R})$. Then, by Theorem 5.7C,

$$\text{Biend } E(\overline{R})_R = \text{End }_\Lambda E(\overline{R}) = Q^r_{max}(\overline{R})$$

is the maximal right quotient ring of $\overline{R}$, which by Corollary 7.3 is a semiprimary Levitzki ring. This proves (2) and (3); (4) follows from Theorem 5.7A(3); and (5) comes from Lemma 7.1A and the proof of Theorem 7.1.

8.5 COROLLARY If R has a faithful Δ-quasi-injective module E then the injective hull E(R) is Δ-injective, and $E^n = E(R) \oplus X$ where the countermodule E is generated by n elements. (Cf. Theorem 10.4.)

Proof: Follows from Theorem 8.4 and its proof.

8.6 THEOREM Any Δ-injective right R-module E is a direct sum $E = \oplus_{i=1}^m H_i$ of finitely many homogeneous Δ-injective modules $\{H_i\}_{i=1}^m$. Moreover, there exist finitely many ideals $A_0, A_1, \ldots, A_n \in A_r(E,R)$ such that every indecomposable direct summand of E is isomorphic to a direct summand of $QE(R/A_i) = E_{R/A_i}(R/A_i)$, $i = 0, \ldots, n$.

Proof: Since E is Σ-injective by Theorem 7.1, then $E = \oplus E_i$ is a direct sum of indecomposable Σ-injectives by Proposition 5.9. Since E is faithful over $\overline{R}_0$, then $E^{n_0} = QE(\overline{R}_0) \oplus X_0$ in mod-R by Theorem 8.5, for $n_0 < \infty$. Then $QE(\overline{R}_0)$ is also Σ-injective, hence also a direct sum $\oplus_{i\in I_0} E_i^0$ of indecomposable Σ-injective modules by Azumaya's unique decomposition theorem. This means that every indecomposable direct summand of E comes from a direct summand of either $QE(\overline{R}_0)$ or X_0. Since $\overline{R}_0$ is finitely generated, then I_0 is a finite set, and we may write $E = G_0 \oplus Y_0$ in mod-R, where each indecomposable direct summand of G_0 is isomorphic to some E_i^0 with $i \in I_0$, and Y_0 has no such direct summand. Y_0 is also Δ-injective, so by Proposition 5.9 again, either $Y_0 = 0$ in which case the theorem is true with n = 0, or else $A_1 = r_E Y_0 \supset A_0$, and $Y_0^{n_1} = QE(\overline{R}_1) \oplus X_1$ where $\overline{R}_1 = R/A_1$. Moreover, by the part just proved, $Y_0 = G_1 \oplus Y_1$ in mod-R with each indecomposable direct summand of G_1 isomorphic to a direct summand of $QE(R_1)$. By Theorem 7.1, the acc in $A_r(E,R)$ holds, and since $A_i \in A_r(E,R)$ for all i, then an induction establishes that $E = G_0 \oplus \cdots \oplus G_n$, where $G_i = E_{\overline{R}_i}(\overline{R}_i)$ is a direct sum of indecomposable modules where $\overline{R}_i = R/A_i$, i = 0, ..., n. Since $QE(\overline{R}_i)$ is a finite direct sum of indecomposable modules, then $E = H_1 \oplus \cdots \oplus H_m$, where H_i is the direct sum of isomorphic indecomposable modules, i = 1, ..., m.

If R is any noetherian ring with an infinite number $\{V_i\}_{i=1}^{\infty}$ of nonisomorphic simple right modules, then $E = \oplus_{i=1}^{\infty} E(V_i)$ is Σ-injective with infinitely many fully invariant direct summands. Thus, the following immediate corollary, obtained independently by J. Golan, does not hold for Σ-injective modules.

8.7 COROLLARY If E is as in Theorem 8.6, then E has just finitely many fully invariant direct summands. If F is one of these, then $F = \oplus_{i=1}^{m} F_i$, where $F_i = H_i$ or $F_i = 0$.

Proof: We prove this and at the same time give another proof of the main statements of Theorem 8.6. As in the proof of Proposition 5.15, $FQ \subseteq Q$ for *any* direct summand F of E in mod-R. Thus, since End E_R = End E_Q, the fully invariant R-direct summands coincide with the fully invariant Q-direct summands of E. Similarly, $E = \oplus_{i\in I} E_i$ for indecomposable injectives $\{E_i\}_{i\in I}$, and E_i is also an indecomposable injective Q-module, for all $i \in I$. (Note: E is injective over Q since E is finendo; hence any R-direct summand of E is injective over Q.) Since Q is semiprimary, then E and E_i are soc modules, so $E_i = E_Q(W_i)$ for a simple Q-module W_i. But there are just finitely many simple Q-modules $V_1, \ldots, V_m$, so there is a partition $I = \coprod_{i=1}^{m} I_i$, where $E_j \approx E_Q(V_i)$, $j \in I_i$, and $H_i = \oplus_{j\in I_i} E_j$ is homogeneous. Let $E_i^0 = E_Q(V_i)$. Then $E_i^0 \approx E_j^0 \Leftrightarrow i = j$.

Now if F is a fully invariant direct summand, then by 6.1, $F = \oplus_{j\in J} F_j$, where F_j is isomorphic to some E_i^0, and in fact, $E_i^0 = uF_j$ for some (unit) $u \in \Lambda = \text{End } E_R$. This proves that $E_i^0 \in H_i \cap F$. But every indecomposable direct summand E_i of H_i is similarly isomorphic to E_i^0, so $H_i \subseteq F$, and hence F has the stated structure. This proves that there are only finitely many of the F's.

Caution: We do not claim that the H_i are fully invariant. (Are they?) However:

8.8 COROLLARY If $Q = \text{End } E_R$ is primary-decomposable, then $H_1, \ldots, H_m$ are the minimal fully invariant direct summands of E.

Proof: Write $Q = Q_1 \times \cdots \times Q_t$ for primary rings $Q_1, \ldots, Q_t$. First assume t = 1, so Q has a unique simple module, and m = 1 in Theorem 8.6 and Corollary 8.7, and hence the Δ-injective module E has no nontrivial fully invariant direct summands. In the general case, clearly $H_i = EQ_i$, i = 1, ..., m, so $\text{Hom}_R(H_i,H_j) = \text{Hom}_Q(H_i,H_j) = 0$ for $i \neq j$; that is, H_i is fully invariant. If $0 \neq F$ is any fully invariant direct summand of H_i, then F contains a submodule E_j isomorphic to E_i^0, and by injectivity, there is an endomorphism f of E mapping E_j onto E_i^0. But $f(E_j) \subseteq F$ since F is fully invariant, so $f(F) = H_i$.

8.9 THEOREM (Converse of the Teply-Miller-Hansen theorem) An injective module E is Δ-injective iff E is Σ-injective and some between ring S is semiprimary. Then $Q = \text{Biend } E_R$ is semiprimary.

Proof: The necessity follows from Theorems 7.1 and 8.3, and the converse apes the proof of corollary 4 of Ref. 1: Let

$$A_1 \supseteq A_2 \supseteq \cdots \supseteq A_n \supseteq \cdots \tag{1}$$

be a chain of right ideals in $A_r(E,R)$. Then

$$\{\ell_E A_i\} \quad \text{is descending} \tag{2}$$

and by Lemma 5.18 there is a chain $\{A_i'\}$ of finitely generated right ideals of R such that $\ell_E A_i = \ell_E A_i'$ for every i. Then

$$A_1'S \supseteq A_2'S \supseteq \cdots \supseteq A_n'S \supseteq \cdots \tag{3}$$

is a descending chain of finitely generated right ideals of S, and $\ell_E A_i'S = \ell_E A_i' = \ell_E A_i$ for all i. However, a semiprimary ring satisfies the dcc on finitely generated right ideals (N.B.), hence the chain (3) terminates at, say, n = N, and therefore so does the chain (2). But $\ell_E A_N = \ell_E A_{N+1} \Rightarrow A_N = A_{N+i}$ for all i, so (1) also terminates at n = N. This proves that E is Δ-injective over R, S, and Q by (1) of Theorem 5.8B.

A right R-module M is *Björkian*, or a Björk module, if M satisfies the dcc on finitely generated submodules. A ring R is *right Björkian* if the canonical right R-module R is.

8.10 THEOREM (Björk [30]) 1. If M satisfies the dcc on principal right ideals, then M is Björkian.

2. If R satisfies the dcc on principal right ideals, then every right R-module is Björkian.

3. A ring R is right Björkian iff every right R-module is Björkian.

8.11 COROLLARY (Bass-Björk) (1) A left perfect ring is right Björkian; in particular, a semiprimary ring is Björkian (both sides).

(2) Moreover, any right Björkian ring satisfying acc ⊥ (i.e., upper Levitzki) is semiprimary.

Proof: (1) follows from Theorem 8.10, and (2) follows from proposition 1, section 4 of Ref. 1. (Hint: Show there exists n such that $J^{n\perp} = R$, where J = rad R, hence $J^n = 0$.)

8.12 COROLLARY Any Σ-injective module over right Björkian ring R is Δ-injective; in fact, any upper Levitzki right R-module is lower Levitzki.

Proof: Same proof as Theorem 8.9 for the case S = R.

8.13 COROLLARY If R is right Björkian, and if R has a faithful upper Levitzki module E, then R, $Q^r_{max}(R)$, and Q = Biend E_R are semiprimary rings. Thus, any right Björkian right Σ-ring (e.g., right artinian ring) R has semiprimary $Q^r_{max}(R)$.

Proof: For then R is upper Levitzki by Proposition 3.2, so Corollary 8.12 applies to conclude that R is semiprimary. Since E is Σ-injective by Corollary 8.12, then E(R) is Δ-injective by Corollary 8.5, and hence $Q^r_{max}(R)$ = Biend $E(R)_R$ and Q are semiprimary by Theorem 8.9.

Trivially any direct sum of copies of a Σ-injective is Σ-injective; but is it obvious that the direct product is Σ-injective? The next result follows easily from the corresponding result for upper (lower) Levitzki modules.

8.14 THEOREM Any direct summand F of the direct sum $E^{(A)}$ or product E^A of a $\Sigma(\Delta)$-injective module E is $\Sigma(\Delta)$-injective.

Proof: Apply Proposition 3.5 to conclude that F has the desired property when F is $E^{(A)}$ or E^A. Moreover, any submodule of an upper (lower) Levitzki module has the same property, and a direct summand of $E^{(A)}$ or E^A is injective.

By Example 5.17 a direct sum of Δ-injectives need not be even Σ-injective, and hence neither is the direct product inasmuch as a submodule of a Levitzki module is Levitzki by Proposition 3.1.

8.15 THEOREM Every injective right R-module is $\Sigma(\Delta)$-injective iff R is right noetherian (artinian).

Proof: The sufficiency is trivial. Conversely, since every right ideal of R belongs to $A_r(E,R)$ for any cogenerating module E (e.g., Ref. 11, p. 184, proposition 23.13), then $\Sigma(\Delta)$-injectivity of the injective hull of E implies that R is right noetherian (artinian).

9. KASCH RINGS

Kasch rings, defined presently, are also called scalar rings or S-rings in the literature. The main theorems on these rings are:

1. $Q = Q^r_{max}(R)$ is Kasch iff $R \hookrightarrow Q$ is a flat epimorphism such that $IQ = Q$ for all dense right ideals of R (Proposition 9.2). (Cf. Ref. 59.)

2. If E is Δ-injective, then Q = Biend E is a right Kasch right artinian ring iff J = rad Q is a right annihilator ideal. (See Theorem 9.5 for this and equivalent conditions.)

3. A sufficient condition for 2. is commutativity of the prime ideals of Q; equivalently, Q is primary-decomposable (Theorem 9.5).

4. Sufficient conditions for Q to be right artinian when E is Δ-injective are:

   (a) J^2 is a right annihilator ideal.

   (b) $J^2 = 0$.

   (c) J/J^2 is finitely generated (Theorem 9.5).

   (d) R has a Kasch overring (Theorem 9.1A).

The concept of a dense right ideal appears in the next result. In general, a submodule N of a module M is *dense* if $\mathrm{Hom}_R(S/N,M) = 0$ for all submodules $S \supseteq N$. Then, by the definition of a rational extension (above Theorem 5.7A), N is dense in M iff M is a rational extension of N.

9.1 PROPOSITION AND DEFINITION A ring A is said to be *right Kasch* if it satisfies any of the following equivalent conditions.

K1. Every simple right A-module embeds in A.

K2. For each maximal right ideal M, $\ell_A M \neq 0$.

K3. A has no dense right ideals $\neq$ A.

K4. Each maximal right ideal of A is an annihilator right ideal.

K5. E(A) cogenerates mod-A.

K6. $M^* = \mathrm{Hom}_A(M,A) \neq 0$ for all nonzero finitely generated $M \in$ mod-A.

K7. Every right ideal of A belongs to $A_r(E,A)$, where $E = E(A)$.

Proof: See, e.g., Ref. 13, p. 235, lemma 5.1, for equivalence of K1 through K5. Also K1 $\Leftrightarrow$ (K6 for M simple). Thus K6 $\Rightarrow$ K1. Conversely, if $0 \neq M$ is finitely generated, then M has a maximal submodule M_1; hence K1 $\Rightarrow$ $\mathrm{Hom}_A(M/M_1,A) \neq 0$, so $M^* \neq 0$.

Now K7 $\Rightarrow$ K2, and K5 $\Rightarrow$ K7, e.g., by Ref. 11, p. 184, proposition 23.13.

Remarks: (1) Any primary decomposable semiprimary ring is Kasch (see Corollary 9.9).

(2) Any right Kasch ring S is right rationally closed (via K3); hence $S = Q^r_{max}(S)$. Thus, if $A \subseteq S \subseteq Q^r_{max}(A)$, then S right Kasch implies that $S = Q^r_{max}(A)$.

9.1A COROLLARY Let E be an injective right R-module, and let S be a right Kasch between ring for E.

(1) Then E is Δ-injective iff S is right artinian.

(2) If E is finendo, then E is Σ-injective iff S is noetherian (cf. Corollary 9.7B).

In either case, $S \approx Q \approx Q^r_{max}(\overline{R}) = Q^r_{max}(S)$.

Proof: By Theorem 5.8B, E is $\Sigma(\Delta)$-injective over R iff it is over S. This proves that S right noetherian (artinian) suffices for E to be $\Sigma(\Delta)$-injective over R. Conversely, any Δ-injective module is finendo, so in either case, E is finendo; hence by Corollary 5.6A, E is injective, and $E^n = E_S(S) \oplus X$ in mod-S for some $n < \infty$. Since S is Kasch, every right ideal of S belongs to $A_r(E_S(S),S)$, so $\Sigma(\Delta)$-injectivity of E over S implies the acc (dcc) in $A_r(E_S(S),S) \subseteq A_r(E,S)$, so S is right noetherian (artinian).

By Theorem 5.7C, there is an embedding $S \hookrightarrow Q \hookrightarrow Q' = Q^r_{max}(S) = Q^r_{max}(\overline{R})$. By the remark following 9.1, $S = Q^r_{max}(S)$; hence $S \approx Q$; canonically.

Since $Q = Q^r_{max}(R) = \text{Biend } E(R)_R$ canonically (Theorem 5.7A), then a between ring S for E(R) may be represented as a subring of Q containing R, and is called an *overring of* R.

9.1B COROLLARY If R has a right Kasch overring, then R is a right $\Sigma(\Delta)$-ring iff S = Q is right noetherian (artinian).

Proof: Since $Q = \text{Biend } E(R)_R$, we can apply Corollary 9.1A to E = E(R) (which is necessarily finendo).

9.2 PROPOSITION The following properties of a ring R are equivalent:

(a) $Q = Q^r_{max}(R)$ is right Kasch.

(b) $R \hookrightarrow Q$ is a flat epimorphism and IQ = Q for a right ideal I of R iff I is dense in R.

Proof: See Ref. 13, p. 236, proposition 5.2.

9.3 PROPOSITION If R is a right Kasch ring, then J = rad R is a right annihilator, and then if R has just finitely many maximal right ideals, J is a finite right annihilator of R.

Conversely, if R/J is right Kasch, e.g., if R/J is semisimple Artin, and if J is a finite right annihilator, then R is right Kasch.

Proof: By Proposition 9.1, each maximal right ideal M is the annihilator $x^\perp$ of an element $x \in R$; hence J, the intersection of maximal right ideals, is a right annihilator, and is a finite right annihilator if the set of maximal right ideals is finite.

Suppose R/J is right Kasch, and $J = r_Q(S)$ for $n = |S| < \infty$. Then, by Proposition 4.1, $R/J \hookrightarrow R^n$. Since every simple right R-module V embeds in R/J, then V embeds in R^n, hence in R, so R is right Kasch.

9.4 COROLLARY A semilocal right Δ-ring R is right Kasch iff J = rad R is a right annihilator.

Proof: One way follows from Proposition 9.3. Conversely, if R is a Δ-ring, then R is Levitzki; hence $A_\ell(R)$ satisfies acc, and then by Proposition 4.3, if J = r(L) is a right annihilator, then $J = r(L_1)$ for a finitely generated left ideal L_1, so R is right Kasch by Proposition 9.3.

Below, we let $\text{soc}_r R$ denote the right socle of R, namely the sum of the simple modules, and $\text{soc}_\ell R$ denote the left socle.

9.5 THEOREM Let E be Δ-injective over R, and let $Q = \text{Biend } E_R$. If Q is right Kasch, then Q is right artinian. Necessary and sufficient conditions for Q to be right Kasch right artinian are the following:

(1) $J = \text{rad } Q \in A_r(E,Q)$.

(2) $J \in A_r(R,R)$.

Moreover, the following four conditions are sufficient for Q to be right Kasch right artinian.

(3) $r_Q(\text{soc}_r Q) = J$.

(4) $J = \mathcal{D}_\ell$, the left singular ideal of Q.

(5) $J = r_Q(\text{soc}_\ell R)$.

(6) $\text{soc}_r Q$ is an essential left ideal.

(7) $\text{soc}_r Q \supseteq \text{soc}_\ell Q$.

Furthermore, the following conditions imply that Q is right artinian:

(8) $J^2 \in A_r(Q,Q)$.

(9) $J^2 \in A_r(E,Q)$.

(10) J/J^2 is finitely generated (examples: $|J/J^2| < \infty$, or Q/J^2 is right Goldie).

(11) R is commutative.

(12) Q is primary-decomposable.

Proof: The first conclusion comes from Corollary 9.1A. Next, for any right ideal I of Q,

$$r_Q(E \cdot \ell_Q I) = r_Q \ell_Q I$$

since E is faithful over Q. Thus, $I \in A_r(E,Q)$ iff $I \in A_r(Q,Q)$, so (1) ⇔ (2). Also, Q right Kasch ⇒ (1), e.g., by K7. Conversely, (1) ⇒ Q right Kasch via, e.g., Proposition 9.3, using the fact that Q is Levitzki to obtain by Proposition 4.3 that every right (left) ideal of Q is a finite annihilator, and the fact that Q is semiprimary, hence semilocal.

(3) ⇒ (2) is trivial.

(2) ⇒ (3) Since Q is semiprimary, then $\ell_Q(J) = \text{soc}_r Q$, so (2) ⇒ (3).

(4) ⇒ (2) and (4) ⇒ (5) Let $\mathcal{D}_r = \text{sing } Q_Q$ and $\mathcal{D}_\ell = \text{sing }_Q Q$. Since Q is semiprimary, Q has left essential $S_\ell = \text{soc}_\ell Q$ and right essential $S_r = \text{soc}_r Q$, so $\mathcal{D}_\ell = r_Q S_\ell$ (and $\mathcal{D}_r = \ell_Q S_r$), i.e (4) ⇒ (2) and (4) ⇒ (5).

(5) ⇒ (4) follows because $\text{soc}_\ell Q$ is left essential in Q.

(6) ⇒ (7) follows since every essential left ideal contains every simple left ideal.

(7) ⇒ (4) Since Q is Levitzki, then $\mathcal{D}_r$ and $\mathcal{D}_\ell$ are nil via acc ⊥ (see, e.g., Ref. 12, p. 394, lemma 9.7), and nilpotent, by a theorem of Levitzki, Herstein, and Small ([11], p. 20, corollary 17.23), hence contained in J. But, if $S_r \supseteq S_\ell$, then

$$J \supseteq \mathcal{D}_\ell = r_Q S_\ell \supseteq r_Q S_r \supseteq J$$

so (4) holds.

This completes the proof of the sufficiency of (4) through (7). (8) ⇒ (9) by the proof of (1) ⇒ (2), so that for (8) it suffices to prove (9).

(9) is proved similarly to (1) of the theorem: $F = r_E J^2$ is Δ-injective, hence Σ-injective, over Q/J, so Proposition 5.12 implies that Q/J^2 is Goldie, and hence has finite right socle. Therefore, J/J^2 has finite length, which suffices to establish that Q/J^2 is right artinian. But in a semiprimary ring this implies Q right Artin. (Cf., e.g., Ref. 11, p. 187, 23.19.)

Another proof of (9) by Nakayama uses the fact that for any primitive idempotent $e \in Q$, eJ/eJ^2 is simple, hence $\approx fQ/fJ$ for some other primitive idempotent $f \in Q$, and then $eJ^2/eJ^3 \approx fJ/fJ^2$ has finite length, so $|eQ/eJ^3| < \infty$. Since J is nilpotent, an obvious induction shows that $|eQ| < \infty$ for all such e, and then $|Q| < \infty$.

(10) follows from the proof of (9).

(11) ⇒ (12) (See, e.g., Theorem 9.8 below.)

(12) ⇒ Q is Kasch via Corollary 9.9 below.

A ring Q has *radical-square-zero* if $J(Q)^2 = 0$.

9.6 COROLLARY If E is Δ-injective over R, then $Q/r_Q J^{n-2}$ is right artinian, where $Q = \text{Biend } E_R$, J = rad Q, and n is the index of nilpotency of J. In particular, if $Q = Q^r_{max}(R/r_R E)$ has radical-square-zero, then Q is right artinian.

Proof: Since E is faithful over Q, then $A = r_Q J^{n-2} = r_Q(EJ^{n-2})$ belongs to $A_r(E,Q)$. Since $Q = Q^r_{max}(R/r_E R)$ is a Δ-ring by Theorem 5.7B, then $\overline{Q} = Q/A$ is a Δ-ring by Corollary 10.3 below. However, $\overline{J} = \text{rad } \overline{Q} = (J + A)/A$ (since Q and $\overline{Q}$ are semiprimary), and $J^2 \subseteq A$; that is, $\overline{J}^2 = 0$, so (8) of Theorem 9.5 implies that $\overline{Q}$ is artinian. In particular, if $J^2 = 0$, then Q is artinian.

9.7 COROLLARY A commutative ring R has a faithful Δ-injective module E iff $Q = Q_{max}(R)$ is artinian. In this case, Biend E_R is artinian, and the injective hull E(Q) [= E(R)] is a Δ-injective cogenerator of mod-Q which is Δ-injective over R. (Cf. Theorem 11.4.4.)

Proof: Biend ${}_RE$ is artinian by Theorem 9.5(11) when E is Δ-injective. By Theorem 8.6, the existence of E implies that E(R) is Δ-injective over R, hence over $Q = Q_{max}(R)$ by Theorem 5.7B. Since R is commutative, Q is artinian by Theorem 9.5(11), and E(Q) cogenerates mod-Q by 9.1 since Q is Kasch.

Conversely, Q artinian implies that every injective Q-module is Δ-injective. Since E(R) = E(Q), we have E(R) is Δ-injective by Theorem 5.7 again.

9.7A COROLLARY Under the same assumptions, $Q_{max}(R) \approx$ Biend ${}_RE$ canonically.

Proof: Since E is finendo, then $E^n = E(R) \oplus X$ in mod-R for some $n < \infty$, and, moreover, S = Biend ${}_RE$ maps E(R) into itself; hence S induces a subring $\overline{S}$ of $Q = Q_{max}(R) = \text{End}\ {}_\Lambda E(R)$, where $\Lambda = \text{End}\ {}_RE(R)$. (Note: Λ is induced by End ${}_RE$.) Since E is faithful, and $R \hookrightarrow S$ canonically, then $S \approx \overline{S}$. Now $\overline{S}$ is artinian; hence $Q_{max}(\overline{S}) = \overline{S} \supseteq Q_{max}(R) = Q$, proving the corollary.

9.7B COROLLARY If E is a faithful finendo injective module, then S = Biend E_R canonically induces a subring $\overline{S} \approx S$ in $Q = Q^r_{max}(R)$. If S is rationally closed, then $S \approx Q^r_{max}(R)$ canonically.

Proof: Same proof as Corollary 9.7A, using the fact that $Q^r_{max}(\overline{S}) = \overline{S}$ if S is rationally closed.

9.8 THEOREM (Asano) For a semiprimary ring Q the following are equivalent:

(1) Q is primary-decomposable.
(2) The prime ideals of Q commute.
(3) The prime ideals of Q are comaximal in pairs.
(4) $R/(\text{rad } R)^2$ is primary-decomposable.

A proof of this may also be found in Ref. 11, p. 50, 18.37.

9.9 COROLLARY Any primary-decomposable ring Q is right and left Kasch

Proof: For any ring decomposition $Q = A \times B$, a simple right Q-module V has the form $V \approx A/K$ for a right ideal of $Q \subseteq A$, or similarly, $V \approx B/K$. This follows because $A = eQ$ for a central idempotent e and $B = fQ$ for $f = 1 - e$, so either $Ve \neq 0$ or $Vf \neq 0$. In the former case $ve \neq 0 \Rightarrow vA = V$ for some $v \neq 0$ in V, and then there is a canonical epimorphism $A \longrightarrow V$ sending $a \longmapsto va$.

Thus, in order to prove Q is Kasch it suffices to show this for each primary direct factor of Q, so suppose $Q = T_n$, where T is completely primary. Now T has a unique maximal ideal J, and hence Q has a unique maximal ideal J_n, and $Q/J_n \approx T/J$ is a simple ring, proving that Q has a unique simple (right or left) module. Since a primary ring has essential socle (right or left), every simple Q-module embeds in Q, so Q is right Kasch.

9.10 PROPOSITION If E is Δ-injective over R, and if $Q = \text{Biend } E_R$ has a unique minimal ideal, then Q is a primary right artinian ring.

Proof: First, if V_1 and V_2 are simple right ideals, then QV_i is a direct sum of modules $\approx V_i$, $i = 1, 2$, and every submodule P of QV_i is a semisimple direct summand of QV_i, $i = 1, 2$. If $P = QV_1 \cap QV_2 \neq 0$, then P is a semisimple right Q-module which is a direct summand of both QV_1 and QV_2; hence $V_1 \approx V_2$. Since Q has essential right socle S_r [e.g., by the fact (Theorem 9.5) that Q is semiprimary], then the minimal ideal S meets S_r, $S \cap S_r \neq 0$, so $S = QV$ for a simple right ideal V. If V_2 is any other simple right ideal of Q, the fact that QV_2 is a minimal ideal yields $QV_2 = QV$, so $V_2 \approx V$.

Similarly, Q has essential left socle, and $WQ = S = QV$ for any simple left ideal W, and therefore Q has just one isomorphism class {V} of simple right ideals, and one {W} of simple left ideals. Moreover, since $S_r = QV = WQ = S_\ell$ and $r_Q J = S_\ell$, then

$$J \subseteq \ell_Q S_\ell = \ell_Q S_r = \mathcal{D}_r \subseteq J$$

by the arguments of Theorem 9.5, where $\mathcal{D}_r$ is the right singular ideal [e.g., see (4) ⇒ (2)]. Since J therefore is an annihilator left ideal, then, by Corollary 9.4, Q is left Kasch; hence every simple left module embeds in Q. Since every simple left ideal of Q is ≈ W this implies that Q/J is simple artinian, that is, that Q is a primary ring. Then Q is right Artin by Theorem 9.5(12).

9.11 COROLLARY If R is a right uniform right Δ-ring, then Q is a local right artinian ring with simple right socle.

Proof: By definition, E = E(R) is Δ-injective. Since E = E(Q), Q is uniform, so S_r is a simple right ideal of Q, and clearly the unique minimal ideal.

For a theorem on right nonsingular right Σ-rings, see Theorem 11.4.9 ff

10. Δ-RINGS

By Corollary 8.16 every injective right R-module is Δ-injective iff R is right artinian. In the obvious way, one can also characterize the condition $Q = Q^r_{max}(R)$ is right artinian, and taking cognizance of when Q is a flat left R-module, or when Q is an epic of R in rings, this can be rephrased entirely within mod-R. See Proposition 10.5 and the corresponding characterization for ring epics (Theorem 10.8).

Aside from this, there is very little new in this chapter, being as it is, merely statements about Δ-injectives specialized to the case E = E(R). Thus: Q is always semiprimary Levitzki and right Goldie (Theorem 10.2), and whatever makes a semiprimary ring artinian gives a theorem about Δ-rings, e.g., Q is right artinian if J/J^2 has finite length; thus if it is Goldie, then Q/J^2 right Goldie suffices. (It follows that Q/J^2 a right Δ-ring suffices.) This happens whenever J^2 belongs to $A_r(E(R),R)$, e.g., when $J^2 = 0$.

Another sufficient condition for right artinian Q is for Q to be right Kasch, e.g., primary-decomposable, which it is when R, whence Q, is commutative. (See Theorem 10.8 ff.)

10.1A THEOREM A ring R is a right Δ-ring iff R has a faithful Δ-injective right R-module E. In this case, $E^n = E(R) \oplus Y$ in mod-R for some integer $n > 0$.

Proof: If E is faithful and Δ-injective over R, then since E is finendo [e.g., by Proposition 8.1(1)], then R embeds in E^n for some $m > 0$ (e.g., Proposition 5.5); hence $E^n = E(R) \oplus Y$ as stated, and E(R) is Δ-injective over R [e.g., Proposition 3.1(2)]. The converse is trivial.

10.1B COROLLARY A commutative ring R is a Δ-ring iff $Q_{max}(R)$ is artinian.

Proof: Apply Propositions 5.7B, 9.7, and 10.1A.

Actually, in this case, $Q_{c\ell}(R)$ is artinian (see Theorem 11.4.4).

10.2 THEOREM (1) Any right Δ-ring R is a right Σ-ring, and R/A is a Δ-ring for any ideal $A \in A_r(E(R),R)$.

(2) Any right Δ-ring R is a right Goldie Levitzki ring.

(3) A ring R is a right Δ-ring iff $Q = Q^r_{max}(R)$ is a right Δ-ring.

(4) If R is a right Δ-ring, then $Q = Q^r_{max}(R)$ is a semiprimary ring.

(5) R is a right Δ-ring iff R is a right Σ-ring and Q is semiprimary.

Proof: (1) R is a Σ-ring by Theorem 7.1. Since $E = E(R/A)$ is a counter-Jordan-Hölder module, A is finitely annihilated; hence by Proposition 4.4, $R/A \hookrightarrow E^n$, for $0 < n < \infty$. Now $F = E_{R/A}(R/A)$ is an essential extension of R/A as an R-module, so $F \hookrightarrow E$. Then F is Δ-injective over R/A by Proposition 3.1(2). (Cf. Theorem 5.19.)

(2) follows from Corollary 3.3 and the Double Annihilator Condition 5.3.

(3) is Proposition 5.7B.

(4) is a special case of Theorem 8.4 applied to Q.

(5) is a special case of Theorem 8.9.

10.3 COROLLARY TO THEOREM 8.4 If R is a right Δ-ring, then:

(1) $E(R) = E_1 \oplus \cdots \oplus E_n$ is a finite direct sum of indecomposable Δ-injective modules E_i, $i = 1, \ldots, n$.

(2) $\Lambda_i = \text{End } E_{i_R}$ is a completely primary ring, and E_i has finite length over Λ_i, $i = 1, \ldots, n$.

(3) $Q_i = \text{Biend } E_{i_R} = Q^r_{max}(R/r_R E_i)$ is a semiprimary right Δ-ring, $i = 1, \ldots, n$.

Proof: (1) follows from Corollary 5.10 and Proposition 3.1(1).

10.4A THEOREM A right Δ-ring R has right artinian $Q = Q^r_{max}(R)$ iff Q is fully right Goldie.

Proof: Any right noetherian ring is fully right Goldie. Conversely, since Q is semiprimary by Corollary 10.3, then Q is right artinian by Corollary 6.6.

10.4B COROLLARY A maximal quotient ring Q is a fully right Δ-ring iff Q is right artinian.

Proof: Any right Δ-ring is right Goldie by Theorem 10.2, so Q is right artinian by Theorem 10.4A.

The next result states that the preservation of injectives by the inclusion functor mod-S $\hookrightarrow$ mod-R defined by a ring homomorphism R $\hookrightarrow$ S is a criterion for flatness.

10.5 PROPOSITION A ring homomorphism R $\longrightarrow$ S defines S as a (left) flat R-module iff every injective S-module is injective as canonical R-module.

Proof: See, e.g., Ref. 12, proposition 6.28.

Recall the convention: "R is a right $\Sigma(\Delta)$-ring" means that R is a right Σ-ring (resp. a right Δ-ring).

10.6 THEOREM Let R $\longrightarrow$ S be a flat homomorphism, and let F be an injective S-module.

(1) If F is Σ-injective over S, then F is Σ-injective over R.

(2) If R $\hookrightarrow$ S is a flat embedding, and if F is Δ-injective over S, then F is Δ-injective over R. Moreover, R is a $\Sigma(\Delta)$-ring if S is.

Proof: (1) is an immediate consequence of Proposition 10.5.

(2) $A_r(F,R) \subseteq A_r(F,S)$ satisfies acc (dcc), so F is Δ-injective over R. Again by Proposition 10.5, the injective hull $E_S(S)$ of S in mod-S contains $E(R) = E_R(R)$, and thus E(R) is $\Sigma(\Delta)$-injective by Theorem 5.19, if S is a right $\Sigma(\Delta)$-ring.

10.7 COROLLARY R is a right $\Sigma(\Delta)$-ring iff R has a flat embedding in a right $\Sigma(\Delta)$-ring.

When R $\longrightarrow$ S is a ring epic [31], then mod-S $\hookrightarrow$ mod-R reflects injectives, and then an S-module which is Σ-injective over R is Σ-injective over S.

10.8 THEOREM For a right Δ-ring R, the following conditions on $Q = Q^r_{max}(R)$ are equivalent:

(1) E = E(R) cogenerates mod-Q.

(2) Q is right Kasch and right Artin.

(3) J = J(Q) is a right annihilator ideal.

(4) J belongs to $A_r(E,Q)$.

(5) Every ideal of Q belongs to $A_r(E,Q)$.

(6) Every right ideal of Q is a right annihilator.

Proof: Immediate from Theorem 9.5 and its proof.

10.9 COROLLARY If R has primary-decomposable $Q = Q^r_{max}(R)$, then R is a right Δ-ring iff Q is right artinian. This happens in any of the following cases:

(a) R is commutative.
(b) R has local $Q^r_{max}(R)$.
(c) R has primary $Q^r_{max}(R)$.
(d) The prime ideals of Q commute.

10.10 COROLLARY If R has primary (e.g., local) $Q = Q^r_{max}(R)$, then R is a right Δ-ring iff Q is right artinian.

Appendix

A ring R has a *Morita duality* iff there is an injective cogenerator U of mod-R which is also an injective cogenerator over $\Lambda = \text{End } E_R$ and $R = \text{End }_\Lambda E$. If X is a right R-module, then $X^* = \text{Hom}_R(X,U)$ is a left Λ-module. Similarly, if Y is a left Λ-module, set $Y^* = \text{Hom}_\Lambda(Y,\Lambda)$. Then X^* (resp. Y^*) is called the U-*dual* of X(Y). Furthermore, then $X^{**} = (X^*)^*$ is defined, and there is a natural transformation of functors $h : 1_{\text{mod-R}} \longrightarrow \pi$ on mod-R under which $h(X) : X \longrightarrow \pi(X) = X^{**}$ is a morphism of mod-R. If h(X) is an equivalence (isomorphism) in mod-R, then X is said to be U-*reflexive*.

Briefly a Morita duality, defined by the module U, is a category duality between the category S_U of U-reflexive right R-modules and the category $_US$ of U-reflexive left Λ-modules induced by the functor $\text{Hom}_R(\ ,U)$, that is, sending $X \longrightarrow X^*$. (See Ref. 11, chapter 23, for further details.)

If $\Lambda \approx R$, then we may suppose $R = \text{End } U_R$, and call the resulting duality a *self-duality of* R.

10.11 EXAMPLES 1. If A is a commutative local noetherian ring, then $U = E(A/\text{rad } A)$ defines a Morita self-duality on $\Lambda = \text{End }_A U$. E.g., if $A = \mathbb{Z}_p$ for some prime number p, then $U = \mathbb{Z}_{p^\infty}$, and $\Lambda = \mathbb{Z}_{(p)}$ is the ring of p-adic integers. Note that $\Lambda = \text{Biend }_R U \neq Q_{max}(\Lambda)$ in this case, since $Q_{max}(\mathbb{Z}_p)$ is a field (= $\mathbb{Q}$, the field of rational numbers).

2. Another self-duality is defined on any algebra R of finite dimensions over a field k by the functor $\text{Hom}_R(\ ,k) \approx \text{Hom}_R(\ ,U)$ where $U = \text{Hom}_R(\ ,R)$. (See Ref. 11, pp. 177-178.)

10.12 PROPOSITION If R has a Morita duality defined by U, and if U is cofaithful (equivalently, finendo), then $R = Q^r_{max}(R)$.

Proof: Apply Corollary 9.7B.

10.13 THEOREM (Osofsky [32]) If R has a Morita duality defined by U, then R is a semiperfect ring. If R has dcc on finitely generated right ideals (i.e., R is left perfect), then R is right artinian and $\Lambda = \text{End } U_R$ is left artinian.

Cf. Ref. 11, pp. 187-188, corollary 23.18 and proposition 23.20.

10.14 COROLLARY If R has a Morita duality induced by a Δ-injective module U, then $R = Q^r_{max}(R)$ is right artinian.

Proof: Since U is finendo by Theorem 7.1, then $R = Q = Q^r_{max}(R)$ by Proposition 10.12, and Q is semiprimary (hence perfect on both sides) by Theorem 8.6, so R is right artinian by Theorem 10.13.

10.15 COROLLARY If R is a right artinian ring with a Morita duality, then $R = Q^r_{max}(R)$.

10.16 COROLLARY If R is any algebra of finite dimensions over a field k, then $R = Q^r_{max}(R)$.

11. INJECTIVE MODULES OVER NONNOETHERIAN COMMUTATIVE RINGS

11.1 Introduction

In this chapter we prove the theorem of Beck [33], essentially (1) of Theorem A (Theorem 11.1.1) which completely determines the structure of Σ-injective modules over nonnoetherian commutative rings. First, consider the concepts:

(1) $P \in \text{Ass } M$ signifies that P is a (prime) ideal of R which is maximal in $A_r(M,R)$. Then P is a *prime ideal associated with* M. Note that Ass M = Ass E(M) and that R/P embeds in M. (Cf. Propositions 11.3.1 and 11.3.2.)

(2) A prime ideal P is called a *noetherian (artinian) prime* if the local ring R_P is noetherian (artinian).

Inasmuch as $E = E(R/P)$ is canonically an R_P-module (Propositions 11.3.1 and 11.3.2), then (2) implies that E is a module over a noetherian ring; in fact, by the corollary to Proposition 11.3.1, E is a module over a noetherian local ring containing R/A, where $A = r_R E$. (See Corollary 11.3.1A.)

11.1.1 THEOREM A Let R be a commutative ring.

(1) (Beck [33]) An injective module E is Σ-injective iff $E = \oplus_{i \in I} E_i$, where for all $i \in I$, $E_i = E(R/P_i)$ for a noetherian prime ideal P_i.

Moreover, for each $i \in I$ the following properties hold:

(2) P_i is the unique prime ideal Q of R such that $E(R/Q) \approx E(R/P_i)$.

(3) E_i cogenerates mod-R_{P_i}.

(4) $E_i = E_{R_{P_i}}(R_{P_i}/PR_{P_i})$ is artinian in mod-R_{P_i}.

Another theorem of Beck characterizes commutative Σ-rings (Σ_0-noetherian rings in Ref. 33 as quotient-noetherian rings. We give a proof of this in Theorem B (Theorem 11.4.2).

Theorems A and B and the structure theory from preceding sections are combined to obtain the structure of Δ-rings and Δ-injective modules over commutative rings.

Thus R is a Δ-ring iff R is quotient-artinian [Theorem C (Theorem 11.4.4)]. This is proved as a consequence of Theorem 10.9 which states that any Δ-injective over R has artinian biendomorphism ring.

Furthermore, an injective R-module E is Δ-injective iff there exist finitely many artinian prime ideals P_i, $i = 1, \ldots, m$, and $E = \oplus_{i=1}^{m} H_i$ where H_i is a fully invariant homogeneous injective submodule isomorphic to the direct sum of modules isomorphic to $E(R/P_i)$, $i = 1, \ldots, m$ (Theorem 11.4.6).

The set of prime ideals $\{P_i\}_{i \in I}$ appearing in Theorem A for the decomposition (1) of E coincides with Ass E, so if E(M) is Σ-injective, then Ass $M \neq \emptyset$.

E is said to be *completely decomposable* if $E = \oplus_{i \in I} E_i$ is a direct sum of indecomposable modules $\{E_i\}_{i \in I}$. Then E will be said to be *homogeneous* provided that $E_i \approx E_j$ for every i and j in I.

11.1.2 COROLLARY A_1 An injective module E is Σ-injective over R iff E is completely decomposable and every associated prime ideal is noetherian.

Once (1) through (3) of Theorem A are proved, (4) is straightforward application of Matlis' theorem [8], and as a corollary one can explicitly determine the endomorphism ring:

11.1.3 COROLLARY A_2 If E is an indecomposable Σ-injective module, then $\Lambda = \text{End}\ _R E$ is the P-adic completion of the local ring R_P, where P is the associated prime ideal.

Let M be a Σ-module. Since $A_r(M,R) \subseteq A_r(E(M),R)$, and the latter satisfies the acc, so does the former, that is, every Σ-module is upper Levitzki. We let Ass M denote the maximal elements of $A_r(M,R)$. [These in fact coincide with Ass E(M); see Corollary A_3.]

11.1.4 COROLLARY A_3 Let M be a Σ-module R.

(1) Ass M = Ass E(M).

(2) Ass M is finite (of cardinal n) iff E(M) is a finite direct sum of (n) indecomposable modules.

(3) A sufficient condition for (2) is for M to have finite Goldie dimension.

In Ref. 34, the elements of Ass M are called *maximal prime ideals* of M, and (3) takes this form: If M is finitely generated and R is noetherian, then Ass M is finite. (See Lemma 11.4.1 in this connection.)

Before going further, we mention that Theorem A also enables one to characterize Δ-injectives in view of the Teply-Miller theorem (Chap. 7). (See Corollary 11.4.3.)

11.2 Preliminaries

For convenient referencing, we list here some of the basic theorems on Σ-injective modules. (Most of these already have been commented on in the text. Consult Ref. 11, chapter 20, for additional references.)

Σ_1: R is right noetherian iff every injective right R-module is Σ-injective.

Σ_2: A module E is Σ-injective iff $E^{(\omega)}$ is injective (Theorem 5.8A).

Σ_3: E is Σ-injective iff E is injective and upper Levitzki (Theorem 5.8A).

Σ_3': E is Σ-injective iff for every right ideal I of R, there corresponds a finitely generated right ideal $I_1 \subseteq I$ such that $\ell_E I_1 = \ell_E I$ (Theorem 5.8A).

Σ_4: If E is injective, then E is Σ-injective iff $E = \oplus_{i\in I} E_i$ is a direct sum of indecomposable Σ-injective modules $\{E_i\}_{i\in I}$ (Theorem 5.9).

Σ_5: R is right noetherian iff R has a Σ-injective cogenerator. Equivalently, if $\{V_j\}_{j\in J}$ is the set of all simple R-modules, then $E(\oplus_{j\in J} V_j)$ is Σ-injective [27]. (See, e.g., Ref. 11, p. 113, corollary 20.3C.)

11.3 Proof of Beck's Theorem

Throughout, R is a commutative ring. When it becomes important to distinguish the injective hull of a module M over a ring $S \neq R$, we let $E_S(M)$ denote the injective hull in mod-S. The next proposition is essentially Matlis's Theorem 3.6 of Ref. 8. (As will be seen, the noetherian hypothesis pervasive in Ref. 8 is not needed here.)

11.3.1 PROPOSITION If P is a prime ideal, then $E = E(R/P)$ is canonically a faithful R_P-module, and $E = E_{R_P}(R_P/PR_P)$ cogenerates mod-R_P.

Proof: We first prove

$$P = \{a \in R \mid \ell_E a \neq 0\} \tag{11.1}$$

and

$$\ell_E P = \{\cap \ell_E a \mid a \in R,\ \ell_E a \neq 0\} \tag{11.1'}$$

For if $0 \neq y \in E$, then there exists $r \in R$ such that $0 \neq yr \in xR$, where $xR \approx R/P$ canonically (i.e., $P = r_R x$). Then we have $yr = xb$ with $b \in R\backslash P$ and $tb \neq 0$, for all $t \in R\backslash P$, so $yt \neq 0$, proving that $P \supseteq \{a \mid \ell_E a \neq 0\}$. However, any $a \in P$ has $\ell_E a \supseteq \ell_E P \neq 0$, so (11.1) holds, and (11.1') follows.

Second,

$$Et = E \quad \forall t \in R\backslash P \tag{11.2}$$

To see this we use the following property of injective modules (e.g., Ref. 11, p. 189, proposition 23.21):

$$Ea = \ell_E r_R a \quad \forall a \in R \tag{11.3}$$

[One can replace a by a finitely generated ideal a in (11.3).] Next let $A = r_R E$. Then E is still injective over $\overline{R} = R/A$, so (11.3) implies that

$$Ea = E\overline{a} = \ell_E r_{\overline{R}} \overline{a} \quad \text{for any coset } \overline{a} = a + A \in \overline{R} \tag{11.4}$$

If $t \in R\backslash P$, then $\ell_E t = 0$ by (11.1); hence if $b \in R\backslash A$, then $bt \notin A$ since $Eba \neq 0$. This proves that $r_{\overline{R}}\overline{t} = 0$, so (11.3) implies $E = Et = E\overline{t}$ as asserted.

Since (11.1) and (11.2) imply that E is divisible and torsion free with respect to $R\backslash P$, then E is canonically an R_P-module. If $F = E_{R_P}(E)$, then F is essential over E as an R_P-module, hence as an R-module (via the canonical homomorphism $R \longrightarrow R_P$), so $F = E$ via injectivity of E in mod-R. Furthermore, $V = R_P/PR_P$ embeds in E since $r_{R_P} x = PR_P$, and hence $V \approx xR_P \subseteq E$, so $E = E_{R_P}(V)$. Since E is injective, and V is the unique simple R_P-module, then E is the (unique) minimal cogenerator of mod-R_P. This of course implies that E is a faithful R_P-module. However, a direct computation is less fancy: If $a \in A$, then (11.3) implies that $\ell_E r_R a = 0$, so $r_R a \subsetneq P$. If $t \in R\backslash P \cap r_R a$, then at $= 0 \Rightarrow a/1 = 0$ in R_P, so $AR_P = r_{R_P} E = 0$.

Let S^* denote the regular elements of a ring S.

11.3.1A COROLLARY Let $A = r_R E(R/P)$, and $\overline{R} = R/A$. Then $\overline{R}\backslash\overline{P} \subseteq \overline{R}^*$, so $\overline{R} \hookrightarrow \overline{R}_{\overline{P}}$ canonically.

Since E is faithful over $\overline{R}$, then E is faithful over $\overline{R}_{\overline{P}}$, so the canonical isomorphism $R_P \approx \overline{R}_{\overline{P}}$ yields another way to see that E is faithful over R_P.

11.3.2 PROPOSITION (1) If M is an R-module and if $A_r(M,R)$ has a maximal element P, then P is a prime ideal, and E(R/P) is an indecomposable direct summand of E(M). Moreover, (2) if $E(R/P) = E(R/Q)$ for an ideal Q, then $P \supseteq \sqrt{Q}$, where

$$\sqrt{Q} = \{x \in R \mid x^n \in Q,\ n \gg 0\}$$

In particular, (3) P is the unique prime ideal P' such that $E(R/P) = E(R/P')$.

Proof: (1) The proof of (1) is standard for noetherian rings, and the noetherian hypothesis is needed only for the existence of P. (See, e.g., Ref. 13, p. 161.) R/P is uniform, so E(R/P) is indecomposable. Since $R/P \approx xR \subseteq M$, then $E(R/P) \subseteq E(M)$. Since E(R/P) is injective, it is a direct summand of every module containing it.

(2) If $y \in \sqrt{Q}$, then $y^n \in Q$, so $\ell_E y \neq 0$; hence $y \in P$ by (11.1).

(3) By (2), $P' \supseteq P = \sqrt{P} \supseteq \sqrt{P'} = P'$.

11.3.3 COROLLARY If $E = E(R/P) = E(R/Q)$ [in Proposition 11.3.2(2)] is Σ-injective, then

(1) $P = \{a \in R \mid \exists\ t \in R\backslash P \text{ with } (at)^n \in Q,\ n \gg 0\}$.

(2) $\cap_{n<\omega} P^n \subseteq A = r_R E$.

Proof: This will use Theorem A and the Krull intersection theorem.
(1) The corollary is true when R is noetherian (e.g., Ref. 13, p. 164, corollary 1.15), so $PR_P = \sqrt{QR_P}$; hence for every $a \in P$ there exist $q \in Q$ and $t \in R/P$ such that $a^n/1 = q/t$, which means that $(a^n t - q)s = 0$ for some $s \in R\backslash P$. By the proof of Proposition 11.3.1, $a^n t - q \in A = r_R E \subseteq Q$, so $a^n t \in Q$, and $(at)^n \in Q$. Conversely, if $(at)^n \in Q$, as stated, then $\ell_E at \neq 0$, so $\ell_E a \neq 0$ whence $a \in P$ by (11.1).
(2) By the Krull intersection theorem, $\cap_{n<\omega} I^n = 0$ for any ideal I in the radical of a noetherian ring; hence for $I = PR_P$, we have $(\cap P^n)R_P = 0$, hence $(\cap P^n)t = 0$ for some $t \in R/P$, proving that $\cap P^n \subseteq A$.

Proof of Theorem A: (1) The acc on $A_r(E,R)$ implies the acc on $A_r(E_i,R)$, so E_i is Σ-injective. Moreover, by Proposition 2.2, for any module M, a right ideal I belongs to $A_r(M,R)$ iff $I/A \in A_r(M,R/A)$ where $A = r_R M$ is the annihilator of M. Since M injective over R implies R injective over R/A (by virtue of the fact that any R/A essential is R-essential), then every E_i is Σ-injective over R/A_i, where $A_i = r_R E_i$. Now any maximal element P of $A_r(E_i,R)$ is a prime ideal of R and has the form $P = r_R x$ for $x \in E_i$; hence $R/P \approx xR$ embeds in E_i, so $E_i = E(R/P)$. We now apply Proposition 11.3.1 to conclude that E_i is canonically an injective R_P-module which cogenerates mod-R_P. Moreover, the same proof shows that E_i is Σ-injective over mod-R_{P_i}, so R_{P_i} is noetherian by Σ_5. This proves the necessity of (1). The sufficiency is a consequence of Σ_1 and Σ_5. Thus, since R_{P_i} is noetherian, E_i is Σ-injective over R_{P_i}; hence E_i is Σ-injective over R by Theorem 10.6, and then Σ_4 applies.

(2) and (3) follow from Propositions 3.1(2), and since R_{P_i} is noetherian, $E_i = \mathrm{Hom}_{R_{P_i}}(R_{P_i},E_i)$ is artinian by Matlis' duality theorem [8].

Proofs of the Corollaries to Theorem A: A_1 and A_2 are immediate. Also $A_3(2)$ and $A_3(3)$ follow immediately from $A_3(1)$. Proof of $A_3(1)$: Since

$M \subseteq E = E(M)$, then $A_r(M,R) \subseteq A_r(E,R)$, and hence every P which is maximal in $A_r(M,R)$ belongs to Ass E. Conversely, let $P \in$ Ass E, so $E_1 = E(R/P)$ embeds in E. By (11.1) and (11.1'), $P = r_R y$ for any $y \neq 0$ in $\ell_{E_1} P$. In particular, choosing $x = yr \neq 0$ in $E_1 \cap R \neq 0$, we get $P = r_R x$ for $x \in R$. Obviously, P is maximal in $A_r(M,R)$, so $P \in$ Ass M.

11.4 Commutative Sigma and Delta Rings

A commutative ring R is said to be *quotient-noetherian (artinian)* provided that $Q_{c\ell}(R)$ is noetherian (artinian).

$B_1 \Leftrightarrow B_2$ of the next theorem is a theorem of Beck [33].

11.4.1 THEOREM B For a commutative ring R the following are equivalent:

B_1: R is quotient-noetherian.

B_2: R is a Σ-ring.

B_2': $A_r(E(R),R)$ satisfies the acc.

B_3: R embeds in a Σ-injective module E.

B_4: R embeds in some finite product E^n of copies of a Σ-injective E.

B_5: There exists a Σ-injective faithful module E which is finitely generated over $\Lambda(E_R) = \text{End } E_R$.

B_6: Ass R is finite and every $P \in$ Ass R is noetherian.

When this is so, then $Q = Q_{c\ell}(R) = Q_{max}(R)$, and E(R) cogenerates mod-Q.

For the proof of Theorem B, we require the next lemma. By Corollary A_3, Ass M is finite iff M has finite Goldie dimension.

11.4.2 LEMMA Let M be an R-module such that Ass M is finite and $E = E(M)$ is faithful. Then an ideal I of R either contains a regular element of R, or else I is contained in some $P \in$ Ass M (equivalently, $\ell_M I \neq 0$).

Proof: Since Ass M = Ass E we may as well take M = E. If I does not contain a regular element, that is, if $\ell_R x \neq 0$ for all $x \in I$, then $\ell_E x \neq 0$ for all $x \in I$, since E is faithful. Therefore, each $x \in I$ belongs to some $P_x \in$ Ass E. Since Ass E is finite, this implies that I is contained in some $P \in$ Ass E ([34], p. 55, theorem 81). In this case, $\ell_E I \supseteq \ell_E P$, and hence $\ell_E I \neq 0$. Conversely, $\ell_E I \neq 0$ iff $\ell_M I \neq 0$ since M is essential in I. Assuming $\ell_M I \neq 0$, then I is contained in any maximal element P of $A_r(M,R)$ containing $r_R \ell_E I$, so $I \subseteq P \in$ Ass M.

Proof of Theorem B: $B_4 \Leftrightarrow B_5$ by Proposition 5.5, $B_2 \Leftrightarrow B_2'$ by Σ_3, and both $B_2 \Leftrightarrow B_3$ and $B_3 \Leftrightarrow B_4$ are trivial. Also $B_6 \Leftrightarrow B_2$ via Theorem A.

$B_1 \Rightarrow B_2$: $Q = \text{End}\,{}_\Lambda E$ is the maximal quotient ring $Q_{max}(R)$ of R by Theorem 5.7A, and Q contains $Q_{c\ell}(R)$ canonically. Also, as an R-module, Q coincides with

$$\overline{R} = r_E \ell_\Lambda R = \{x \in E \mid \lambda x = 0,\ \forall\lambda \in \Lambda \text{ with } \lambda(1) = 0\}$$

(Theorem 5.7A). From this, and rationality of $\overline{R} = Q$ over R, one deduces that E is the injective hull of Q in mod-Q.

This holds even if R is noncommutative, and when R is commutative, it is true that $Q = Q_{c\ell}(R)$ if and only if every dense (i.e., faithful) ideal of R contains a regular element [e.g., when $Q_{c\ell}(R)$ is noetherian (see example 4 of Ref. 13, p. 237)]. When this is so, then every simple Q-module embeds in Q, hence in E, inasmuch as then $\ell_Q M \neq 0$ for any ideal $M \neq Q$. Therefore, in this case, E cogenerates mod-Q.

$B_2 \Rightarrow B_1$: Write $E = E(R) = \oplus_{i \in I} E(R/P_i)$ as in Theorem A(1). Then, R being finitely generated (cyclic), essential in E implies that I is a finite set, that is, Ass E(R) = Ass R is finite. Then Lemma 11.4.2 yields that every dense ideal M of R contains a regular element, since $\ell_R M = 0 \Leftrightarrow \ell_E M = 0$. As stated above, this implies that $Q = Q_{c\ell}$ and that E cogenerates mod-Q.

Since $R \hookrightarrow Q$ is epic, by Proposition 10.5, a Q-module F is injective in mod-Q if F is injective in mod-R. Thus, since $E^{(a)}$ is injective in mod-R for any cardinal a, then E is Σ-injective in mod-Q, so Q is noetherian by Σ_5.

$B_1 \Rightarrow B_2$: $Q = Q_{c\ell}$ noetherian implies that $E^{(a)}$ is injective in mod-Q for any a, hence injective in mod-R.

A theorem of Camillo [35] states that a fully Goldie commutative ring is noetherian, where *Goldie* means the following two properties hold in R: acc $\perp$ and acc $\oplus$. Then R is *fully (right) Goldie* if every factor ring of R is (right) Goldie. Analogously, one says R is a *fully* Σ-ring if every factor ring is a Σ-ring, and similarly for fully Δ-rings.

Following Vámos, we say R is *fractionally noetherian (artinian)* iff $Q_{c\ell}(R/I)$ is noetherian (artinian) for all ideals I of R.

11.4.3 COROLLARY For commutative ring R the following are equivalent: (1) R is noetherian; (2) R is a fully Σ-ring: (3) R is fractionally noetherian.

Proof: (3) implies that R is fully Goldie by Theorem 11.4.1, so apply Camillo's theorem to obtain (3) ⇒ (1). But (2) ⇔ (3) by Theorem B and (1) ⇒ (2) is trivial.

The next theorem is "dual" to Theorem B.

11.4.4 THEOREM C For a commutative ring R the following are equivalent:

C_1: R is a Δ-ring.

C_2: R is quotient-artinian.

C_3: There exists a faithful Δ-injective module.

Then this is so, then $Q_{c\ell}(R) = Q_{max}(R)$.

Proof: $C_1 \Leftrightarrow C_3$ as in the proof of Theorem B for faithful Σ-injectives, and, similarly, one proves that $Q = Q_{c\ell}(R)$ is a Δ-ring iff R is a Δ-ring in the same way as was proved in Theorem B for Σ-rings. If Q is a Δ-ring, then Q is a Σ-ring by Theorem 7.1, so Q is noetherian by Theorem B; hence as stated, $Q = Q_{max}(R)$, and $Q_{max}(R)$ is semiprimary by Corollary 7.3 (or Theorem 8.4). Therefore, Q is artinian by the Hopkins-Levitzki theorem (e.g., Ref. 11, 18.12 ff), or by Corollary 6.6.

$C_2 \Rightarrow C_1$: Over an Artin ring, every injective module is Δ-injective; hence $Q = Q_{c\ell}(R)$ is Δ-Levitzki. Since $A_r(E(Q),R) \subseteq A_r(E(Q),Q)$, then the dcc on the latter implies that on the former, so $E(Q) = E(R)$ is Δ-injective over R, proving C_1.

$C_2 \Rightarrow C_3$ is trivial.

$C_3 \Rightarrow C_1$ follows from the fact (demonstrated in the proof of $C_1 \Rightarrow C_2$) that E is finitely generated over Λ, so by Theorem B, E(R) embeds in E^n, with $n < \infty$. Then E(R) is Δ-injective along with E, and the proof is complete.

11.4.5 COROLLARY For a commutative ring R the following are equivalent: (1) R is artinian; (2) R is a fully Δ-ring; (3) R is fractionally artinian.

Proof: Immediate from Theorem C and Corollary 11.4.3. Thus (2) ⇔ (3) by Theorem C, and (1) ⇒ (3) obviously. [Also (1) ⇒ (2) by Theorem 5.7B.] Conversely, (2) ⇒ R is noetherian by Corollary 11.4.3. Furthermore, if P is any prime ideal, then $\overline{R} = R/P$ is fractionally artinian. Thus if $Q \supseteq P$ is a maximal ideal, then $R_Q \approx \overline{R}_{\overline{Q}}$ is a domain which embeds in $Q_{c\ell}(\overline{R})$ canonically, hence is artinian iff $\overline{R}_{\overline{Q}}$ is a field, and then $\overline{R}_{\overline{Q}} = Q_{c\ell}(\overline{R})$; proving that Q = P. Thus, every prime ideal of R is maximal, so R is noetherian by a theorem of I. S. Cohen [36].

11.4.6 THEOREM A module E over a commutative ring R is Δ-injective iff $E = \oplus_{i=1}^{m} H_i$ is a finite direct sum of homogeneous fully invariant injective modules $H_1, \ldots, H_m$, where H_i is a direct sum of modules $\approx E(R/P_i)$ for an artinian prime ideal P_i, $i = 1, \ldots, m$.

Proof: By Theorem 9.5(11), $Q = \text{Biend}\ {}_R E$ is commutative artinian, hence primary-decomposable, so the necessity follows from Theorem 8.6, Corollary 8.8, Theorem A, and Theorem B.

Conversely, assume $m = 1$. By Proposition 11.3.1, $E(R/P_1)$ is canonically an R_{P_1}-module, hence is Δ-injective, and also so is H_1, since R_{P_1} is artinian and flat (see Theorem 10.6). By Corollary 7.5, H_1 is counter-Jordan-Hölder. Let $\Lambda_i = \text{End}\ {}_R H_i$, $i = 1, \ldots, m$, and $\Lambda = \text{End}\ {}_R E$. Since H_i is fully invariant, then $\Lambda \approx \prod_{i=1}^{n} \Lambda_i$ canonically, and so E is counter-Jordan-Hölder, and therefore Δ-injective by Corollary 7.5.

R is said to be *homogeneously Goldie*, or H-Goldie, provided that $E(R) \approx E^n$ for a natural number n and an indecomposable module E. [Thus E(R) is both Goldie and homogeneous.]

11.4.7 THEOREM Let R be a commutative H-Goldie ring. Then the following are equivalent:

(1) R is a Σ-ring.

(2) R is a Δ-ring.

(3) $Q_{c\ell}(R)$ is artinian.

(4) $Q_{max}(R)$ is artinian.

When this is so, then $Q_{c\ell}(R) = Q_{max}(R)$.

Proof: Every Δ-injective is Σ-injective [Theorem 8.1(2)], so (2) $\Rightarrow$ (1). Also, $Q_{c\ell}(R)$ noetherian $\Rightarrow Q_{c\ell}(R) = Q_{max}(R)$, so (3) $\Rightarrow$ (4). Now (4) $\Rightarrow$ (2) by Theorem 5.7B, so (1) $\Rightarrow$ (3) remains to be proved.

(1) $\Rightarrow$ (3) Write $F = E(R) = E^n$ for an indecomposable Σ-injective module E. By Theorem B, E is finendo, say, $E = \Lambda x_1 + \cdots + \Lambda x_n$ with $x_i \in E$, $i = 1, \ldots, n$. For any i, let $I_i = r_R x_i$, and then $R_i = R/I_i \approx x_i R$ embeds in E. Now $F_i = \ell_E I_i \supseteq x_i R$ is Σ-injective over R/I_i by Proposition 5.11A, so the injective hull G_i of R/I_i in mod-R/I_i is contained in F_i. But I_i annihilates G_i, so $G_i = F_i$. Let $\Lambda_i = \text{End}\ {}_R F_i$. By Theorem B,

$$Q_i = \text{End}\ {}_{\Lambda_i} R_i = Q_{c\ell}(R_i) = Q_{max}(R_i)$$

is noetherian. Also Q_i is semiprimary by Theorem 10.2(4), so Q_i is artinian. Since R_i is uniform, the set H of regular elements of R_i is a prime ideal, so $Q_i = (R_i)_H = Q_{c\ell}(R_i)$ is a local ring. Therefore uniformity of Q_i implies simple socle, so Q_i is self-injective, that is, quasi-Frobenius (QF). This implies that $Q_i = F_i$ is the injective hull of R_i, and therefore $\Lambda_i =$ End ${Q_i}_R = Q_i$ canonically. In particular, Λ_i is commutative (and QF). The fact that F_i is quasi-injective as an R-module and $E(F_i) = E$ in mod-R implies that F_i is fully invariant in E by Proposition 5.1A, and therefore $\Lambda_i \approx \Lambda/K_i$ canonically, where $K_i = \ell_\Lambda F_i$ for every i. Moreover, since $K = \bigcap_{i=1}^n K_i$ annihilates E, then K = 0, and hence the canonical homomorphism $\Lambda \longrightarrow \prod_{i=1}^n \Lambda_i$ is an embedding of rings, proving that Λ is commutative.

Now, canonically,

$$Q = Q_{max}(R) = \text{Biend}\ _R E(R) = \text{Biend}\ _R E^n = \text{Biend}\ _R E = \text{End}\ _\Lambda E$$

(see Theorem 5.7C), and commutativity of Λ implies that $\Lambda = Q$ canonically, so Q is local as claimed. By Theorem B, $Q_{c\ell}(R)$ is noetherian, and $Q = Q_{c\ell}(R)$. Since Λ is therefore noetherian, $J(\Lambda) = \Lambda\lambda_1 + \cdots + \Lambda\lambda_m$, with $\lambda_i \in J(\Lambda)$, $i = 1, \ldots, m < \infty$; hence uniformity of E implies that $r_E J(\Lambda) = \bigcap_{i=1}^m \ker \lambda_i \neq 0$. Therefore, there exists $0 \neq x \in E$ such that $xJ = 0$, where $J = J(Q)$, and hence $Q/J \approx xR \hookrightarrow E$, and hence $E = E(Q/J)$ over Q, proving that E cogenerates mod-Q. Moreover, $F = E(Q) \approx E^n$ has essential socle, and hence Q has essential socle, proving that Q is artinian. This completes the proof.

There exist commutative noetherian rings which are not quotient-artinian [37]. The last theorem shows that such rings are not H-Goldie.

The next theorem is a partial converse of the Teply-Miller theorem for an indecomposable injective module over a commutative ring.

11.4.8 THEOREM For an indecomposable injective module E over a commutative ring R, the following are equivalent:

(1) E is Δ-injective.

(2) E is finendo and Σ-injective.

(3) E = E(R/P) for an artinian prime ideal P.

(4) $Q = \text{Biend}\ _R E$ is artinian.

When this is so, then $Q = Q_{c\ell}(R/r_R E) = Q_{max}(R/r_R E)$, and $R/r_R E$ is a Δ-ring. For any Δ-injective E, the direct sum $E^{(I)}$ is Δ-injective, for any set I.

Proof: (1) ⇒ (2) by Theorems 7.1 and 8.1 (without assuming that E is indecomposable).

(2) ⇒ (4) Let $A = r_R E$. Then Proposition 5.5 implies that $R/A \hookrightarrow E^n$, and hence, since E is an indecomposable injective R/A-module, $E_{R/A}(R/A) \approx E^m$ for $m \leqslant n$. Since R/A is thereby an H-Goldie Σ-ring, and

$$Q = Q_{max}(R/A) = \text{Biend }_R E^m = \text{Biend }_R E$$

(as noted in Theorem 5.7C), then by Theorem 11.4.7, Q is artinian local ring.

(4) ⇒ (1) by use if Theorem 5.8B.

Finally, (3) ⇔ (1) by Theorem C.

For the last statements, see Theorem 11.4.4, also Proposition 5.15 and Theorem 5.8B.*

The example of a noetherian commutative ring which is not quotient-artinian shows that the theorem cannot be extended to nonhomogeneous Goldie E, since E(R) in that example satisfies (2), and Q in (4) is $Q_{c\ell}(R)$.

11.4.9 THEOREM [1] Let R be a right nonsingular ring. Then the following are equivalent:

(1) R is a right Σ-ring.

(2) R is a right Δ-ring.

(3) $Q = Q^r_{max}(R)$ is artinian.

Proof: (3) ⇔ (2) by Theorem 5.7B, and (2) ⇒ (1) by Theorem 10.2(1).

(1) ⇒ (3) follows from the fact that Q is von Neumann regular when R is right nonsingular. Now (1) implies that Q is a Σ-ring (via Theorem 5.7B), so Q is right Goldie by Corollary 5.13.

Beck [33] proved the next result with the conclusion that R is a Σ-ring.

11.4.10 COROLLARY If there exist finitely many prime ideals $P_1, \ldots, P_n$ of R with zero intersection, then R is a Δ-ring, so $Q_{c\ell}(R)$ is artinian.

Proof: In a commutative ring, a nilpotent element lies in every prime ideal, so R is reduced (i.e., has no nilpotent elements except 0), hence

---

*Namely $Q = \text{Biend }_R E^{(I)}$ by Proposition 5.15, Q is artinian by Theorem 9.5(11), and $E^{(I)}$ is Δ by Theorem 5.8B.

nonsingular (i.e., no element except 0 has essential annihilator ideal), so the maximal quotient ring $Q = Q_{max}(R)$ is a von Neumann regular ring, and $Q = E(R)$ (e.g., Ref. 11, 19.34 ff., or Ref. 23). Since every essentially closed (i.e., complement) ideal of Q is a direct summand, these all belong to $A_r(Q) = A_r(E,Q)$. It follows that an ideal I of R is a complement ideal iff $I = J \cap R$ with $J \in A_r(E,Q)$, that is, iff $I \in A_r(E,R)$. An easy exercise shows that $A_r(E,R)$ is the set $A_r(R)$ of annihilator ideals of R (or refer to Utumi [38]). Now $Q_{c\ell}(R/P_i)$ is a field, $i = 1, \ldots, n$; hence R embeds in a semisimple artinian ring, namely, $R \hookrightarrow \Pi_{i=1}^n Q_{c\ell}(R/P_i)$ canonically. Now any subring of a ring with dcc $\perp$ has the same property, so $A_r(E,R) = A_r(R)$ satisfies dcc, so R is a Δ-ring, and therefore $Q_{c\ell}(R)$ is artinian by Theorem 11.4.4.

It is clear from Theorem 11.4.4 that $Q_{c\ell}(R) = Q_{max}(R)$ in Corollary 11.4.10.

11.4.11 COROLLARY TO 11.4.9 For an indecomposable injective E over a commutative ring R the following are equivalent:

(1) E is Δ-injective.

(2) E^n is Σ-injective and finendo for some integer $n \geq 1$.

(3) E^n is Σ-injective and finendo for every integer $n \geq 1$.

(4) $E^{(A)}$ is Δ-injective for any index set A.

(5) E^A is Δ-injective for any index set A.

(6) E^A is Σ-injective for any index set A, and E is finendo.

Proof: If E is finendo, so is E^n, $E^{(A)}$, or E^A as an application of Proposition 5.5 readily shows, since if $\overline{R} = R/r_R E$, then one of these modules is cofaithful over $\overline{R}$ iff all of them are. In particular, (2) $\Leftrightarrow$ (3), so by Theorem 11.4.8 we have (3) $\Leftrightarrow$ (1).

11.4.12 PROPOSITION If E is indecomposable and injective over a commutative ring R, then E is Σ-injective iff $Q = \text{Biend}\ _R E$ is noetherian. If $E = E(R/P)$, then $Q = R_{(P)}$, the P-adic completion of R_P, for the prime ideal P.

Proof: The sufficiency holds for any injective module over any ring, by Theorem 5.8B(2). By Theorem A, if $E = E(R/P)$ is Σ-injective, then P is a noetherian prime, and then $R_{(P)} = \text{End}\ _R E$ by Matlis's theorem (cf. Corollary A_2). Then $Q = R_{(P)}$ trivially.

Remark: A general Σ-injective module E need not have noetherian Q, as the example $E = \mathbb{Q}/\mathbb{Z}$ demonstrates, since B = End E is the product of all of the p-adic rings $\mathbb{Z}_{(p)}$, one for each prime number p, and then Q = B.

11.4.13 PROPOSITION Over a commutative ring R, a homogeneous injective R-module $H = E^{(I)}$, where E is indecomposable, is Σ-injective iff Q = Biend $_RE$ is noetherian (and iff S = Biend $_RH$ is noetherian). Then Q = S.

Proof: Follows from Proposition 5.15 (stating that Q = S), and the Cailleau-Beck theorem.

11.5 *Artinian (Noetherian) Injectives Are Sigma (Delta) Injective*

The theorem stated in the section heading is for commutative R.

11.5.1 THEOREM If E is an artinian (noetherian) injective module over R, then E is Σ(Δ)-injective. Moreover B = End $_RE$ is noetherian (artinian, and so is Q = Biend $_RE$). The converse holds for indecomposable injectives.

Proof: E is counterartinian (-noetherian), so Corollary 5.11D applies, and B is noetherian (artinian) by Proposition 6.8 (6.7).

The converse follows from Theorems A, B, and 11.4.6 inter alia. Thus, E = E(R/P) for P a noetherian prime, and then $B = R_{(P)}$, the P-adic completion of R_P, is noetherian when E is Σ-injective and indecomposable. If E is Δ-injective, then E is Σ-injective, and endowed as stated, and, moreover, Biend $_RE = R_{(P)}$ is artinian; hence $B = R_{(P)} = R_P$ is artinian.

11.5.2 COROLLARY If E is noetherian injective over R, then Q = Biend $_RE$ is artinian, and E defines a Morita duality between the category of E-reflexive Q-modules and that of E-reflexive modules over B = End $_RE$.

Proof: Since E is noetherian over R, then E is noetherian over Q, and E is Δ-injective by Theorem 11.5.1, so Q is artinian and E(Q) cogenerates mod-Q by Theorem 9.5(11). Since $E^n = E(Q) \oplus X$ in mod-Q for finite n, by Theorem 8.4, then E^n, hence E, contains a copy of each simple Q-module, and therefore E also generates mod-Q. Since E is noetherian over Q, then E is a finitely generated injective cogenerator, hence defines a Morita duality over the artinian ring Q. (See Ref. 39, or Ref. 11, chapter 24.)

11.6 Δ-Polynomial Rings Are Polynomial Rings Over Δ-Rings

The theorem stated in the title is the Δ-ring case of the next theorem, whereas the Σ-ring case is Beck's theorem.

11.6.1 THEOREM If R is commutative, then the polynomial ring $R[X_1,\ldots,X_n]$ in finitely many variables over R is a Σ(Δ)-ring iff R is.

Proof: By Theorem 11.4.1, if R is a Σ-ring, then $Q = Q_{c\ell}(R)$ is noetherian. Since Q[X] is a between ring for R[X], in fact, $Q_{c\ell}(R[X]) = Q_{c\ell}(Q[X])$, and since $Q_{c\ell}$ for a noetherian commutative ring is noetherian, we have that $Q_{c\ell}(R[X])$ is noetherian, so R[X] is a Σ-ring. For the other case, it is known that if a ring R has right artinian $Q_{c\ell}$, then the same is true of R[X] (theorem of Small [40]). This has been generalized to infinitely many variables by Pillay [41]), so the next corollary is proved in the same way.

11.6.2 COROLLARY (Pillay) If R is commutative, then the infinite polynomial ring over R is a Δ-ring iff R is.

PROBLEMS

1. In corollary 4.9 of Ref. 33, Beck proves that a commutative polynomial ring $R[X_1,\ldots,X_n]$ is a Σ-ring (of "noetherian depth" d) iff the same is true for R, but leaves a number of interesting questions, including whether or not the above theorem holds for infinitely many variables, or for R noncommutative.
2. If E is Σ-injective over R and Q = Biend E_R, does there exist a right noetherian (artinian) between ring? Cf. Ref. 42; a sequel to this, same title "II", in preprint, states the answer is "no".
3. If E is counterartinian and injective over R, then E is Σ-injective over R since then, by Proposition 3.1, E is upper Levitzki over R and Theorem 5.8A(3) applies. When does the converse hold? This happens, e.g., when E is Δ-injective since then E is counter-Jordan-Hölder (Theorem 7.1). Also if E is indecomposable injective over a commutative ring R, for then E is an artinian R_P module for some prime P, and $R_P \hookrightarrow \Lambda$ = End E canonically, so E is counterartinian. [See Theorem A, Sec. 11.1, especially (4).] By Proposition 5.15, a Σ-injective module is not necessarily counterartinian.
4. What is the structure of an artinian (noetherian) Σ(Δ)-module? If E is finitely generated over R, then E Σ-injective implies that Λ = End E_R

is right perfect (Proposition 6.3); and E Δ-injective implies that Λ is left artinian (Corollary 6.4).

5. What is the structure of an artinian (noetherian) injective module E over noncommutative R? (For commutative R, refer to Theorem 11.5.1 and Proposition 11.4.2.) By Proposition 6.5 it is known that B = End E_R is semiprimary (left noetherian semiperfect).

6. What is the structure of a ring R such that $A_r(E,R)$ satisfies the dcc (acc) on, e.g., principal right annihilators, or on finitely annihilated right ideals? (Consider the right ideals $r_R X$ for an element X of E, or for X a finite subset of E. Of course, the dcc on the latter implies that E is Δ-injective via the Double Annihilator Theorem 5.3; for commutative R, the acc on the former suffices to yield E = E(R/P) for a prime ideal P whenever E is indecomposable. See Chap. 11.)

7. Related to Problem 1 is the problem of determining right Σ-rings ideal-theoretically, as in the case of R commutative (essentially by saying that R has the acc on closed right ideals, that is, that $Q_{c\ell}(R)$ is noetherian).

8. Theorem 8.9 reduces the structure of a Δ-injective module E to that of a Σ-injective over a semiprimary right Goldie Levitzki ring Q. What are the Σ-injective modules over such a ring? And for this, it suffices to know the indecomposable ones.

9. More generally, what is Hansen's theorem for a Σ-injective module E, that is, what is a good statement about Biend E_R besides that it is right Goldie and upper Levitzki? Even in case R is commutative, the problem is open, except when E is indecomposable or homogeneous (e.g., see 11.4.4).

10. Relative to the false proposition stated in the last paragraph of Chap. 1, it would be interesting to determine all rings over which every noetherian right module is artinian (and the dual problem). Right Björkian rings are trivial examples, since every right module satisfies the dcc on finitely generated submodules. Moreover, do these rings include the left self-injective regular rings? If so, then the proof of Theorem 7.1 can be shortened, since then in the proof we have $|J^i E/J^{i+1}E| < \infty$ as a module over Λ, which is what is needed.

11. Little is known about $Q_{max}(R[X])$ in the general case. If $Q_{max}(R)$ is injective, then P. Pillay has asked (orally) if $Q_{max}(R[X])$ is injective. If R is nonsingular, then "yes", since R[X] is nonsingular, hence has regular self-injective Q_{max}.

12. Along the lines for noetherian commutative R, one might ask if in general is it true that $Q_{max}(R) = Q_{c\ell}(R)$ implies the same for the polynomial

ring R[X]? Nastasescu, and independently, Pillay proved this true when R is nonsingular (i.e., reduced).

13. Storrer's theorem to the effect that if R is right and left artinian, then Q is left artinian ought to fit into the scheme that we have developed, especially in Sec. 10, but this remains open. Also, is Q necessarily two-sided artinian in this case? (Question of Storrer [43].)

NOTES

Over a right artinian ring, of course, any injective right R-module is Δ-injective, but there are other finiteness questions that can be raised, and have been raised. Rosenberg and Zelinsky [44] studied the question of when a module M of finite length $|M|$ has injective hull E(M) of finite length; equivalently, when every simple right R-module V has $|E(V)| < \infty$. They characterized these properties and showed that the existence of a division ring D and a division subring S such that D has unequal right and left dimensions over S is sufficient to construct an artinian ring with simple module V having E(V) of infinite length. Such a division ring was constructed by P. M. Cohn [45,46] in answer to a question of E. Artin posed at the Princeton University Bicentennial celebration in 1947. (Some have questioned the proof of the construction in Ref. 45, so also check in Ref. 46.)

Many papers on $\Sigma(\Delta)$-injective modules by Nastasescu [47-53] have appeared since the writing of these notes, and very briefly I comment on these below, following a resumé which he kindly supplied at my request.

If Q is an injective right R-module, then the set

$$F_Q = \{I \text{ a right ideal} \mid \mathrm{Hom}(R/I,Q) = 0\}$$

is an additive (or Gabriel) topology on R. The class

$$\mathcal{T}_Q = \{M \in \text{mod-}R \mid \mathrm{Hom}(M,Q) = 0\}$$

of modules is the torsion class associated to the additive topology F_Q (it is a localizing subcategory according to Gabriel [54]).

Let T_Q : mod-R $\longrightarrow$ mod-R/$\mathcal{T}_Q$ be the canonical functor. The ring R is said to be F_Q-noetherian (resp. F_Q-artinian) if the object $T_Q(R_R)$ is noetherian (resp. artinian) in the quotient category mod-R/$\mathcal{T}_Q$ (see, for example, Refs. 55, 45, and 56).

It is easy to see that Q is $\Sigma(\Delta)$-injective iff R is F_Q-noetherian (resp. artinian) (see, for example, Ref. 55 or Ref. 45).

The question

Does Δ-injective $\Rightarrow$ Σ-injective? (*)

was raised by Albu and Nastasescu in Ref. 56 (problem 4.8, p. 175) in an equivalent form, and solved for R commutative.

Let F be an additive topology on R if R is F-artinian, then is R is F-noetherian? (**)

(The general noncommutative case is the Teply-Miller theorem [2].)

In Ref. 47, (**) is considered in the context of Grothendieck categories. M is *U-strongly generated* if each subobject M' of M is *U-generated*, i.e., there exists an epimorphism $U^{(I)} \longrightarrow M' \longrightarrow 0$.

THEOREM [47,49] Let U be an artinian object of the Grothendieck category C. If M is an artinian object of C which is U-strongly generated, then M is noetherian.

COROLLARY (General Hopkins-Levitzki theorem for a Grothendieck category C having an artinian generator) If M is an artinian object of C, then M is noetherian (i.e., an object of finite length).

A more elementary proof of the Hopkins-Levitzki theorem in a Grothendieck category based only on the length of a Jordan-Hölder composition series appears in Ref. 49. The general Hopkins-Levitzki theorem has applications to modules; among these is the study of relative finiteness with respect to a projective module (Ref. 47, theorem 2.7, p. 753).

The next theorem is related to results in Chap. 8, e.g., Theorems 8.4 through 8.8, and also to theorems in Chap. 11 on Δ-injectives.

THEOREM Let Q be a nonzero Δ-injective right R-module. Then

(1) There exists a finite number of prime ideals $P_1, P_2, \ldots, P_n$ of R such that Ass $Q = \{P_1, P_2, \ldots, P_n\}$. Moreover R/P_i is a right Goldie ring and $E_R(R/P_i) = I_{P_i}^{n_i}$ for each $1 \leqslant i \leqslant n$, where I_{P_i} is an indecomposable injective R-module and $n_i > 0$.

(2) $Q = \oplus_{\alpha\in A} Q_\alpha$, where Q is an indecomposable injective R-module for each $\alpha \in A$. Moreover, for each $\alpha \in A$, Q_α is isomorphic to an injective of the family $I_{P_1}, I_{P_2}, \ldots, I_{P_n}$.

(3) In case R has the property for each prime ideal P that R/P is a

Goldie ring (for example, when R is a right noetherian ring or R is a P.I. ring), then the ideals $P_1, P_2, \ldots, P_n$ are minimal.

Δ-injective noetherian modules can be characterized as follows, answering a part of Problem 4, and generalizing the corresponding results for commutative R proved in Chap. 11.

THEOREM [48] If Q_R is a noetherian Δ-injective R-module, then:

(1) R/Ann Q is a right artinian ring.

(2) Q_R is artinian.

(3) Biend Q_R is a right artinian ring.

(4) End Q_R is a left artinian ring.

In Refs. 50 and 51 the concept of Δ-injective module is closely connected to the concept of *order* in an artinian ring, and a new version of Small's theorem is given.

THEOREM [50] The following two properties of a ring R are equivalent:

1. R is a right order in a right artinian ring.
2. (a) E_R(R/rad R) is a Σ-injective or a Δ-injective module.

    (b) R satisfies the regularity condition with respect to the ideal rad R.

The theory of Δ-injective modules over the polynomial ring R[X] also appears in Ref. 50, and over rings with Krull dimension in Ref. 51.

The general Hopkins-Levitzki theorem and the Gabriel-Popescu theorem [57] are used in proving the following:

THEOREM [53] A Grothendieck category that has an artinian generator U is equivalent to a module category mod-A, where A is a right artinian ring. (Actually, A ≈ End $_C$U.)

There are relative versions with respect to a Gabriel topology of some corresponding results of Faith and Walker [58] in Ref. 52. The last theorem, amazingly, shows that the Hopkins-Levitzki theorem for rings implies the general Hopkins-Levitzki theorem stated above, since if U is artinian, so is A, etc.!

ACKNOWLEDGMENTS

This article was completed during my visit at the Institute for Advanced Study, 1977-1978. I wish to thank the Faculty for this great privilege, and the Staff for much help. In addition, I was helped by members on various problems. Of those that I fleeced trade-secrets off of, Tony Geramita and Chuck Weibel are remembered for letting me know that they had some left! I am endebted to Professor Daniel Gorenstein, Chairman of the Department of Mathematics of Rutgers University, for arranging time off for my visit, and to Dr. Poobhalin Pillay, visiting scholar at Rutgers, for vigorous readings of the manuscript that led to numerous improvements.

I read this paper in various places in various stages of completion, so I have many apologies to make, and thanks to give for my patient auditors: Professor Hans Brungs (University of Alberta), Professors Eldon and Joan Dyer (City University of New York), Professor Bruno Mueller (McMaster University), Professor John Beachy (University of Northern Illinois). I also delivered a "seminar" talk on the paper at the Antwerp Ring Theory Conference 1978, where I learned of references [42]-[45] (mentioned in the abstract), and this stimulated me to recall the converse of the Teply-Miller Hansen theorem as being in essence a proposition in Ref. 1. (See Theorem 8.9 of the text.) (I suppose no one doubts the value of a lecture to the lecturer!) My final thanks go to Professor Freddy Van Oystaeyen for arranging the Antwerp Ring Theory Conference in that lovely city with the unforgettable carillion feast at the Cathedral (each Monday night), and to the entire staff, particularly J. van Geel and his family for illustrating so well the world-famous Belgian art of enjoying life.

REFERENCES

1. C. Faith, Ring with ascending condition on annihilators, *Nagoya Math. J. 27* (1966), 179-191.

2. M. L. Teply and R. W. Miller, The descending chain condition relative to a torsion theory, *Pac. J. Math. 83* (1979), 207-219.

3. W. Schelter and L. W. Small, Some pathological rings of quotients, *J. Lond. Math. Soc. 2*, 14 (1976), 200-202.

4. G. Hansen, Ph.D. Thesis, Indiana Univ., Bloomington, 1974.

5. M. Harada and T. Ishii, *Osaka J. Math. 9* (1972), 217-223.

6. R. E. Johnson and E. T. Wong, Quasi-injective modules and irreducible rings, *J. Lond. Math. Soc. 36* (1961), 260-268.

7. J. M. Goursaud and J. Vallette, Sur l'envelope des anneaux de groupes reguliers, *Bull. Math. Soc. France 103* (1975), 91-103.

8. E. Matlis, Injective modules over Noetherian rings, *Pac. J. Math. 8* (1958), 511-528.

9. J. Golan, Rings having a composition series with respect to a torsion theory, Preprint, Math. Dept., Haifa U., Haifa, 1978.

10. C. Nastasescu, Modules Σ-injectifs, Preprint, Fac. Math., Univ. de Bucarest, Romania, 1979.

11. C. Faith, *Algebra II: Ring Theory*, Springer-Verlag, Berlin-Heidelberg New York, 1976.

12. ______, *Algebra I: Rings, Modules, and Categories*, Grundlehren der Math. Wiss., Bd. 190, Springer-Verlag, Berlin-Heidelberg-New York, corrected reprint 1981.

13. B. Stenstrom, *Rings of Quotients*, Springer-Verlag, Berlin-Heidelberg-New York, 1976.

14. J. A. Beachy and W. D. Blair, Rings when faithful left ideals are cofaithful, *Pac. J. Math. 58* (1975), 1-13.

15. R. Baer, Abelian groups that are direct summands of every containing abelian group, *Bull. Amer. Math. Soc. 46* (1940), 800-806.

16. B. Eckmann and A. Schopf, Über Injective Moduln, *Archiv. Math. 4* (1953), 75-78.

17. C. Faith and Y. Utumi, Quasi-injective modules and their endomorphism rings, *Archiv. der Math. 15* (1964), 166-174.

18. M. Harada, On quasi-injective modules with chain condition over a commutative ring, *Osaka J. Math. 9* (1972), 421-426.

19. J. Lambek, Localization at epimorphisms and quasi-injectives, *J. Algebra 38* (1976), 163-181.

20. N. Jacobson, *Structure of Rings*, revised, Colloquium Publication, Vol. 37, Amer. Math. Soc., Providence, 1955, 1964.

21. Y. Utumi, On quotient rings, *Osaka Math. J. 8* (1956), 1-18.

22. C. Faith, Modules finite over endomorphism rings, in *Proceedings of the Tulane University Symposium in Ring Theory, New Orleans, 1971*, Lecture Notes in Math. Vol. 246, Springer-Verlag, New York-Berlin-Heidelberg, 1972, 145-190.

23. J. Lambek, *Rings and Modules*, second edition, Chelsea, New York, 1976.

24. J. Lambek, On Utumi's ring of quotients, *Can. J. Math. 15* (1963), 363-370.

25. A. Cailleau, Une caracterisation des modules Σ-injectifs, *C. R. Acad. Sci. Paris 269* (1969), 997-999.

26. K. R. Fuller, On direct representations of quasi-injectives and quasi-projectives, *Atchiv. der Math. 20* (1969), 495-502.

26a. J. H. Cozzens, Homological properties of the ring of differential polynomials, *Bull. Amer. Math. Soc. 76* (1970), 75-79.

27. R. P. Kurshan, Rings whose cyclic modules have finitely generated socles, *J. Algebra 15* (1970), 376-386.

28. G. Azumaya, Corrections and supplements to my paper concerning Krull-Remak-Schmidt's theorem, *Nagoya Math. J. 1* (1950), 117-124.

29. I. Kaplansky, Topological representations of algebras, II, *Trans. Amer. Math. Soc. 68* (1950), 62-75.

30. J-E. Bjork, Rings satisfying the minimum condition on principal ideals, *J. reine u. angew. Math. 236* (1969), 466-88.

31. L. Silver, Noncommutative localizations and applications, *J. Algebra 7* (1967), 44-76.

32. B. L. Osofsky, A generalization of quasi-Frobenius rings, *J. Algebra 4* (1966), 373-387.

33. I. Beck, Σ-injective modules, *J. Algebra 21* (1972), 232-249.

34. I. Kaplansky, *Commutative Rings*, Allyn and Bacon, Inc., Boston, 1970.

35. V. P. Camillo, Commutative rings whose quotients are Goldie, *Glasgow Math. J. 16* (1975), 32-33.

36. I. S. Cohen, Commutative rings with restricted minimum condition, *Duke Math. J. 17* (1950), 27-42.

37. T. O. Talintyre, Quotient rings of rings with maximal condition for right ideals, *J. Lond. Math. Sco. 38* (1963), 439-450.

38. Y. Utumi, On rings of which any one-sided quotient ring is two-sided, *Proc. Amer. Math. Soc. 14* (1963), 141-147.

39. K. Morita, Duality for modules and its applications to the theory of rings with minimum condition, *Sci. Reports, Tokyo Kyoiku Daigaku 6* (1958), 83-142.

40. L. Small, Orders in Artinian rings, II, *J. Algebra 9* (1968), 266-273.

41. P. Pillay, Ph.D. Thesis, U. Witwatersand, Johannesburg, 1978.

42. C. Nastasescu, Conditions de finitude pour les modules, *Revue Roum. de Math. Pures and Appl. XXIV* (1979), 745-58.

43. H. H. Storrer, Rings of quotients of perfect rings, *Math. Z. 122* (1971), 151-165.

44. A. Rosenberg and D. Zelinski, Finiteness of the injective hull, *Math. Z. 70* (1959), 372-380.

45. P. M. Cohn, Embedding rings in skew-fields, *Proc. Lond. Math. Soc. 11* (1961), 511-530.

46. ______, *Skew Field Constructions*, Lecture Notes, London Math. Soc. 1977.

47. C. Nastasescu, Conditions de finitude pour les modules, *Revue Roum. Math. Pures et Appl. 24* (1979), 745-758.

48. ______, Conditions de finitude pour les modules II, *Revue Roum. Math. Pures et Appl. 25* (1980), 615-630.

49. ______, Théorème de Hopkins pour les catégories de Grothendieck, in *Ring Theory, Proceedings of the 1980 Antwerp Conference*, Lecture Notes in Math., Vol. 825, pp. 88-93, Springer-Verlag, Berlin-Heidelberg-New York, 1980.

50. ______, Anneaux de quotients et modules Σ(Δ)-injectifs, *Comm. in Algebra 9* (1981) (to appear).

51. ______, Modules Δ-injectifs sur les anneaux a dimension de Krull, *Comm. in Algebra* (to appear).

52. ______, Modules injectifs de type fini par rapport a une topologie additive, *Comm. in Algebra 9* (1981) (to appear).

53. ______, Quelques observations sur les modules Δ-injectifs (to appear).

54. P. Gabriel, Des catégories abéliennes, *Bull. Soc. Math. France 90* (1962), 323-448.

55. C. Nastasescu, La structure des modules par rapport á une topologie additive, *Tôhoku Math. J. 26* (1974), 173-201.

56. T. Albu and C. Nastasescu, Décompositions primaires dans les catégories de Grothendieck commutatives II, *J. reine und angew. Math. 282* (1976), 172-185.

57. P. Gabriel and N. Popescu, Caracterisations des catégories abéliennes avec générateurs et limites inductives exactes, *C. R. Acad. Sci. Paris 258* (1964), 4188-4190.

58. C. Faith and E. A. Walker, Direct sum representations of injective modules, *J. Algebra 5* (1967), 203-221.

59. E. P. Armendariz and G. R. MacDonald, Maximal quotient rings and S-rings, *Can. J. Math. 24* (1972), 835-850.

Supplementary Reading

G. Azumaya, A duality theory for injective modules, *Amer. J. Math. 81* (1959), 249-278.

C. Faith, *Lectures on Injective Modules and Quotient Rings*, Lecture Notes in Mathematics, Vol. 49, Springer-Verlag, Berlin-Heidelberg-New York, 1967.

C. Faith, Injective Quotient Rings of Commutative Rings, in *Module Theory, Selected Papers and Problems from the Special Session, Seattle, 1977*, Lecture Notes in Math., Vol. 700, Springer-Verlag, Berlin-Heidelberg-New York, 1979.

A. W. Goldie, Semiprime rings with maximum condition, *Proc. London Math. Soc. X* (1960, 2) 1-220.

S. Page, FPF rings and some conjectures of C. Faith, *Bull. Canad. Math. Soc.* (to appear).

B. Sarath and K. Varadarajan, Divisibility of direct sums in torsion theories, *Can. J. Math. 28* (1976), 211-214.

J. Soublin, Anneaux coherents, *C. R. Acad. Sci. Paris Ser. A-B* (1968), A183-6.

T. D. Talintyre, Quotient rings with minimum condition on right ideals, *J. Lond. Math. Soc. 41* (1966), 141-144.

Part II

INJECTIVE QUOTIENT RINGS OF COMMUTATIVE RINGS

ABSTRACT R will denote a commutative associative ring with unit and mod-R the category of all R-modules. Then R is FPF (finite pseudo-Frobenius) if every finitely generated faithful R-module generates mod-R. CFPF (completely FPF) means that every factor ring of R is FPF. FPF rings include Prufer rings and self-injective rings; QF (quasi-Frobenius) rings, Dedekind rings, and almost maximal valuation rings (AMVR) are CFPF. A reduced FPF ring is semihereditary with injective $Q_{c\ell}(R)$, and conversely for semihereditary rings. A local ring R is FPF iff the zero divisors are a waist P such that R/P is a valuation ring (VR) and R_P [= $Q_{c\ell}(R)$] is self-injective [1].

We say R is *essentially upper FPF* if every finitely generated module between the injective hull E(R) of R and R generates mod-R.

1.3 THEOREM If R is essentially upper FPF, then R is quotient-injective, that is, $Q_{c\ell}(R) = Q_{max}(R) = E(R)$.

1.6 THEOREM Every CFPF ring is *fractionally self-injective* (FSI) in the sense of Vámos.

These results verify conjectures of Ref. 1. Moreover, using the structure theorems of Vámos, one shows conversely that every FSI ring is CFPF. Thus, the CFPF rings are the finite products of indecomposable rings of the three types: (1) AMVR; (2) almost maximal h-local domains; (3) almost maximal torch rings. (Almost maximal means that R_P is an AMVR for every maximal ideal P. See Theorem B of Ref. 3.)

Any product of FPF (commutative) rings is FPF [2], whereas Vámos' theorem shows that an infinite product of CFPF rings can never be CFPF. A lemma in the proof of Theorem 1.3 states that E(R) and $Q_{c\ell}(R)$ have the same singular submodule; also that R is integrally closed in Q_{max}. In particular, then R contains all nilpotent (idempotent) elements of Q_{max}. Thus Q = $Q_1 \times Q_2$ for rings iff $R = R_1 \times R_2$, where $R_i = Q_i \cap R$, i = 1, 2. (Actually, R pre-FPF suffices for integral closure. N.B.)

A ring R is *pre-FPF* provided that every finitely generated faithful ideal generates mod-R. As shown in Ref. 1 this is equivalent to the condition that every finitely generated faithful ideal is projective.

5.10 THEOREM R is FPF iff R is quotient-injective and pre-FPF.

Together with Theorem 1.1 this proves:

COROLLARY R is FPF iff R is essentially upper FPF and pre-FPF.

[It is the same thing to require that every finitely generated faithful module M be projective when $E(R) \supseteq M \supseteq R$ or $R \supseteq M$.] Thus the FPF rings are precisely those quotient-injective R in which every finitely generated faithful ideal is invertible [in $Q_{c\ell}(R)$].

1. INTRODUCTION

Let R be a commutative ring with unit, and mod-R the category of all R-modules.

Then R is *FPF* if every finitely generated faithful R-module generates mod-R. These rings include Prufer domains, hence Dedekind domains, almost maximal valuation rings, and self-injective rings (see Ref. 1). Furthermore, any product of FPF (commutative) rings is FPF (see Ref. 2).

If R is FPF, then so is the classical quotient ring $Q_c = Q_{c\ell}(R)$ (see Ref. 1). We say that a ring R is *quotient-injective* if Q_c is injective. In this case Q_c coincides with $Q = Q_{max}(R)$.

1.1 THEOREM A Every commutative FPF ring R is quotient-injective.

This verifies a conjecture stated in Ref. 1, where the conjecture was verified for noetherian, local, or reduced R.

Actually the full force of FPF is not needed for the conclusion of Theorem 1.1. Since R is FPF implies Q_c is FPF, then it suffices to assume the latter since $Q_{c\ell}(Q_c) = Q_c$. Thus:

1.2 COROLLARY R is self-injective iff R is an FPF ring in which every regular element is a unit.

Since $Q_{c\ell}(Q) = Q$ this also characterizes the condition Q self-injective. Also, since commutative self-injective rings are FPF, by a theorem of Utumi we can conclude that Q_c/J is FPF when R is FPF, where J = rad Q.

As stated, the full force of FPF is not needed for the conclusion of Theorem 1.1. If $M \supseteq R$ is an R-module, then M is said to be *upper FPF* over R if every finitely generated submodule of M containing R generates mod-R. Then, if the injective hull E of R is upper FPF, we say that R is *essentially upper FPF*.

1.3 THEOREM B If R is essentially upper FPF, then R is quotient-injective.

By a result proved in Ref. 1, if R is essentially upper FPF, so is Q_c (also Q). (See Proposition 2.9.) Furthermore,

1.4 THEOREM If Q_c (resp. Q) is essentially upper FPF, then Q_c (resp. Q) is injective.

Since R is commutative, every finitely generated projective faithful R-module generates mod-R by a theorem of Azumaya; hence we have the corollary:

1.5 COROLLARY If every finitely generated between module of E over R is projective, then R is quotient-injective.

In Chap. 5 we also prove

5.1 THEOREM R is FPF iff R is quotient-injective and pre-FPF.

$\Leftarrow$ follows since pre-FPF implies classical upper FPF.

In Chap. 6 we take up fractionally self-injective rings. A ring R is *fractionally self-injective* (FSI) if every factor ring of R is quotient-injective. In Ref. 3, Vámos classified the FSI rings as the finite products types of indecomposable rings presently defined.

A domain R is (Matlis) *h-local* if every nonzero prime is contained in a unique maximal ideal and every nonzero ideal is contained in just finitely many maximal ideals. A ring R is *(almost) maximal* in case R_P is AMVR for all maximal ideals P. R is a *torch ring* if (i) R has at least two maximal ideals; (ii) R has a unique minimal prime $P \neq 0$ and P is uniserial waist; (iii) R/P is an h-local domain.

Theorem B of Ref. 3 states that every FSI ring R is a finite product of almost maximal rings $R_1, \ldots, R_n$, where R_i is either a VR, an h-local domain, or a torch ring, $i = 1, \ldots, n$.

A ring R is CFPF if every factor ring is FPF. A corollary of Theorem 1.1 is the following verification of another conjecture in Ref. 1.

1.6 THEOREM R is CFPF iff R is FSI.

The FSI rings arise in the classification of FGC (finitely generated modules are direct sums of cyclic modules) rings (called σ-cyclic in Ref. 1). (See Brandal [4], Vámos [3,5], or R. and S. Wiegand [6].) Let a *Bezout* ring mean a ring in which finitely generated ideals are principal. Then R is FGC iff R is an FSI Bezout ring (Vámos [3]).

1.7 COROLLARY R is FGC iff R a CFPF Bezout ring.

1.8 COROLLARY The following conditions on a local ring R are equivalent (1) R is FGC; (2) R is AMVR; (3) R is FSI; (4) R is CFPF.

(2) ⇔ (4) is specifically Theorem 5B of Ref. 1 and (1) ⇔ (2) ⇔ (3) is Lemma 5 of Ref. 3.

1.9 COROLLARY A domain R is FPF iff R is Prufer; and R is CFPF iff R is an almost maximal h-local (Prufer) domain.

Thus, a domain R is FPF iff R_M is a VR for all maximal ideals M; and R is CFPF iff R is h-local and R_M is an AMVR for all such M.

If R is FPF, then the nil radical N(R) coincides with that of Q (Corollary 2.8; cf. Theorem 7.1). We study FPF rings R such that $Q = Q_{c\ell}(R)$ is semilocal (Chap. 7), PF (Chap. 8), or QF (Chap. 7). These are all finite products $\Pi_{i=1}^n R_i$ of indecomposable rings R_i which are either Prufer or have essential nil radicals $N(R_i) = N(Q_i)$, $i = 1, \ldots, n$.

In Chap. 5 we encounter sandwich rings. R is a *sandwich ring* if $J = \text{rad } Q_c \subseteq R$. Thus, trivially every ring R with semiprimitive Q_c is a sandwich ring since then $J = 0$. [This happens whenever Q_c is a subdirect product (or sum) of fields.] A local ring R is FPF iff R is a quotient-injective sandwich ring such that R/J is a VR (Theorem 5.10). Also, an FPF ring R is a sandwich ring whenever $P = J \cap R$ is a nil ideal (Proposition 5.9). In general, FPF implies that R/P is semihereditary FPF (Theorem 5.5), and conversely when R is a quotient-injective sandwich ring (Corollary 5.7).

2. SURVEY OF RELEVANT BACKGROUND

We first fix notation.

$E(M)$ = the injective hull of an R-module M

$E = E(R)$, the injective hull of R in mod-R

$\Lambda = \text{End}_R E$, the endomorphism ring of E

$J = \text{rad}\,\Lambda$, the Jacobson radical of Λ

If $N \subseteq M$ are R-modules, then the notation $N \underset{ess}{\subset} M$ (or $M \underset{ess}{\supset} N$) means that N is an essential submodule of M.

If $X \subseteq M$, then $r_R X = \{r \mid xr = 0, \forall x \in X\}$ is an ideal of R.

The set $\text{sing}\, M = \{x \in M \mid r_R x \underset{ess}{\subset} R\}$ is the singular submodule of M. It is a fully invariant submodule of M; that is, $f(M) \subseteq M$ for every endomorphism f of M. This implies that sing A is an ideal of A for any ring A, the right singular ideal.

For properties outlined below, refer to Ref. 7 or Ref. 8 (especially chapter 19).

If $M \supseteq S \supseteq N$ are R-modules, then S is called a *between module of* M over N. If $\text{Hom}_R(S/N,M) = 0$ for every between module S, then M is said to be a *rational extension of* N, and N is said to be a *dense submodule of* M. This is equivalent to the condition: for all x, $y \in M$ and $y \neq 0$, there exist $r \in R$ with $xr \in N$ and $yr \neq 0$. Each module N has up to isomorphism a unique maximal rational extension $Q_{max}(N)$, characterizable as the between module of E(N) over N consisting of all $x \in E(N)$ such that $\lambda x = 0$ for every endomorphism λ of E(N) with $\ker \lambda \supseteq N$. (Thus, every dense submodule is essential.)

An important property of $Q_{max}(N)$ is that it contains every rational extension of N in E(N), and furthermore, if N is a dense submodule of M, then $Q_{max}(M) = Q_{max}(N)$. (The concept of a rational extension is due to Lambek and Findlay. See Lambek [7] for this and the following, or refer to Ref. 8, chapter 19.)

The special case when $N = R$ yields $Q_{max}(R)$ as an R-module isomorphic to $\text{End}\,_{\Lambda}E$, and hence $Q_{max}(R)$ is a ring, called the *maximal quotient ring of* R. We list some properties of $Q = Q_{max}(R)$.

(A) Q is commutative and $Q = Q_{max}(Q)$.

(B) $E = E(Q)$, the injective hull of Q both in mod-R and mod-Q.

(C) $Q \supseteq Q_{c\ell}(R) = \{ab^{-1} \mid a,b \in R,\ b \text{ regular}\}$. We abbreviate $Q_{c\ell}(R)$ by Q_c.

(D) $\Lambda = \text{End}\,_{Q}E$, that is, every R-endomorphism of E is a Q-endomorphism.

(E) $E = \Lambda 1 = \{\lambda(1) \mid \lambda \in \Lambda\}$ is a cyclic Λ-module isomorphic to Λ/Ω, where $\Omega = \{\omega \in \Lambda \mid \omega(1) = 0\} = \{\omega \in \Lambda \mid \omega(Q) = 0\}$.

(F) $J = \{\lambda \in \Lambda \mid E \underset{ess}{\supset} \ker \lambda\} = \text{rad}\ \Lambda \supseteq \Omega$.

(G) Λ/J is a self-injective regular ring (see Example 2.4).

(H) If A is a right self-injective ring, then sing A = rad A, the Jacobson radical of A; moreover, A/rad A is right self-injective.

(F) through (H) are theorems of Utumi [9,10]. Osofsky [11] proved that Λ/J is right self-injective.

2.1 THEOREM An R-submodule M of Q is dense iff M is faithful. When this is so, then:

(1) $B = \text{End}\ M_R \approx {}_{Q}(M{:}M) = \{q \in Q \mid qM \subseteq M\}$.

(2) $M^* = \text{Hom}_R(M,R) \approx {}_{Q}(M{:}R) = \{q \in Q \mid qM \subseteq R\}$.

This is theorem 1A, corollary, and theorem 1B of Ref. 1.

When (1) [resp. (2)] holds, we say B (resp. M^*) is *induced by* Q.

2.2 THEOREM If P is a generator over a commutative ring R, then P is finitely generated projective over R iff $\text{End}\,_{R}P$ is commutative. In this case $\text{End}\,_{R}P \approx R$ canonically. In particular, any generator P embedded in $Q = Q_{max}(R)$ is finitely generated projective.

This is essentially theorem 1C of Ref. 1. [However, see Remark 2.3(2) below.] A module $P \hookrightarrow Q$ is said to be *rank-1*.

2.3 REMARKS 1. R is then Morita equivalent to $\text{End}\,_{R}P$; hence

$$R \approx \text{center End}\,_{R}P = \text{End}\,_{R}P$$

as stated.

2. If $P \subseteq Q$ generates mod-R, then P is faithful, hence End $_R P$ is induced by Q, so End $_R P$ is commutative. Thus, the first part of Theorem 2.2 implies the last: any rank-1 generator P is finitely generated projective. This can be proved directly as follows: there exist $f_1, \ldots, f_n \in P^*$, $x_1, \ldots, x_n \in P$, such that $\Sigma_{i=1}^n f_i(x_1) = 1$. Since P is dense in Q, then each f_i is induced by an element of $\Lambda = \text{End } E_R = \text{End } E_Q$, hence for all $p \in P$,

$$p = \sum_{i=1}^{n} f_i(x_i p) = \sum_{i=1}^{n} x_i f_i(p)$$

So P is finitely generated projective by the dual basis lemma.

The next result is a counterexample to an assertion made in the last section of Ref. 1.

2.4 EXAMPLE Λ/J need not be commutative.

Proof: Let $A = k[x_1,\ldots,x_t]$ be the polynomial ring in $t > 1$ variables over a field k, let $M = (x_1,\ldots,x_t)$, and $R = A/M^2$. Then

$$N = M/M^2 \approx (R/N)^t$$

is an essential ideal of R, hence $E = E(R) = E(R/N)^t$, so that $\Lambda = B_t$, the $t \times t$ matrix ring over $B = \text{End } E(R/N)_R$. Since $t > 1$, Λ is noncommutative.

(Pre-)Upper FPF Rings

If M is a module, an *overmodule* P of M is a module $\supseteq M$. If M is an essential submodule, then P will be called an essential extension, or essential overmodule of M. Of course, then P is a between module of E(M) over M. If R is FPF, then each finitely generated overmodule of R generates mod-R. When every finitely generated (essential) overmodule of R [i.e., a finitely generated between module of E(R) and R] generates mod-R, then we say that R is (essentially) upper FPF. If every finitely generated overmodule of R contained in Q (resp. Q_c) generates mod-R, then we say that R is *upper-quotient FPF* or *pre-upper FPF (upper-classical quotient FPF*, or *classically pre-upper FPF)*. This definition derives from the definition (given in Ref. 1) for a pre-FPF ring R, namely, a ring such that every finitely generated faithful ideal generates mod-R.

2.5 THEOREM R is pre-FPF iff every finitely generated faithful ideal I is projective, and R is (classically) pre-upper FPF iff every finitely generated between module M of Q over R (Q_c over R) is finitely generated projective. Furthermore, then I (resp. M) is invertible in Q, that is, $I^{-1}I = R$, where $I^{-1} = (I:R)_Q \approx I^*$ [resp. $M^{-1}M = R$, where $M^{-1} = (M:R)_Q \approx M^*$].

Proof: Apply Theorems 2.1 and 2.2.

In a domain R, the R-modules contained in $Q = Q_c$ are called "ideals" whereas those contained in R are "integral" ideals. The invertible ideals of R, then, are nonzero R-submodules of Q which are generators (i.e., finitely generated projective).

The analogy of FPF rings with Prufer domains is made more specific by the next result, inasmuch as a domain is Prufer iff semihereditary. Also, essentially upper FPF rings are integrally closed in Q. (See Proposition 2.7.)

2.6A THEOREM [1] R is a reduced FPF ring iff R is a semihereditary and $R \hookrightarrow Q$ is a flat-epic.

Proof: See proposition 3B and theorem 4 of Ref. 1.

Since a (von Neumann) regular ring is its own quotient ring, we have:

2.6B COROLLARY A regular ring R is FPF iff R is self-injective.

An element x of an overring S is *(weakly) integral* over R if the subring R[x] generated by x and R is finitely generated (or contained in a finitely generated R-submodule). R is (W-)*integrally closed* in S if every (weakly) integral element of S belongs to R.

2.7 PROPOSITION If R is (classical) upper-quotient FPF then R is integrally closed in Q (in Q_c). (Compare Corollary 3.10.)

Proof: Let $x \in Q$ be integral over R. Since $M = R[x]$ is a finitely generated faithful R-module, then M generates mod-R, and by Theorem 2.2, $\text{End}\ _RM \approx R$ canonically. Since M is a ring, then $M \subseteq \text{End}\ _RM$, so $M = R$, and $x \in R$.

2.8 COROLLARY Under the same assumption, R contains every idempotent (nilpotent) element of Q (of Q_c).

Proof: Trivial in view of Proposition 2.7.

If Q_2 is any ring between R and Q, then Q_2 is an R-module, and as such is an essential extension of R. This shows that as a Q_2-module, E is an essential of Q_2, and moreover, the injective hull F of Q_2 in mod-Q_2 coincides with E, since $F \supseteq E$ and F is an essential extension of R in mod-R. (The proof uses the fact that Q_2 is a rational extension of R: if $0 \neq y \in F$ and $q \in Q_2$ is such that $0 \neq yq \in Q_2$, then there exists $r \in R$ such that $qr \in R$ and $yqr \neq 0$. Thus $yR \cap Q \neq 0$, hence $yR \cap R \neq 0$.) A similar proof shows that $Q_{max}(Q_2) = Q$, and that sing Q_2 is the same considering Q_2 as an R-module or as a Q_2-module.

2.9 PROPOSITION If R is (classically) upper-quotient FPF, then so is any between ring Q_2 of Q over R (of Q_c over R). Similarly for R [(essentially) upper] FPF.

Proof: Theorem C of Ref. 1 states that FPF implies that Q_2 is FPF, and the same proof works here. The main point is that if $M = x_1Q_2 + \cdots + x_nQ_2$ is faithful over Q_2, then $M_1 = x_1R + \cdots + x_nR$ is faithful over R, since $R \subseteq Q_2$. Moreover, $R \hookrightarrow Q_2$ is a flat-epic (in the category of rings) since any finitely generated between R-module of Q_2 over R generates mod-R, hence is finitely generated projective over R. (See proposition 1B of Ref. 1.) Also, if M is a between module of Q and Q_2, then M_1 is a between module of Q and R; hence M_1 generates mod-R, and therefore $M \approx M_1 \otimes_R Q_2$ generated mod-Q_2 ([1], proposition 1B).

3. LEMMAS

In this chapter we prove several lemmas needed in the proof of Theorem B (Theorem 1.3) and sufficient to prove that Q is injective. The first proposition is known.

3.1A PROPOSITION If M generates mod-R and if M is finitely generated, say $M = \Sigma_1^n x_iR$, then there exist $f_1, \ldots, f_n \in M^* = \mathrm{Hom}_R(M,R)$ such that $\Sigma_1^n f_i(x_i) = 1$, and hence there is an epic $M^n \approx R \oplus X$ in mod-R.

Proof: Since M generates mod-R, $\Sigma_1^m g_i(M) = R$ for some $g_1, \ldots, g_m \in M^*$, $m < \infty$, so $\Sigma_1^m g_i(y_i) = 1$ for some $y_i \in M$, $i = 1, \ldots, m$. Write $y_i = \Sigma_1^n x_jr_{ij}$ for some $r_{ij} \in R$, $i = 1, \ldots, m$, $j = 1, \ldots, n$. Then the elements $f_j = \Sigma_{i=1}^m g_ir_{ij} \in M^*$, $j = 1, \ldots, n$, have the stated property.

3.1B COROLLARY If M generates mod-R, then $M^* = \text{Hom}_R(M,R)$ generates mod-R.

Proof: In the notation of Proposition 3.1A, $(M^*)^n \approx R \oplus X^*$. So M^* generates mod-R.

3.2 PROPOSITION If $\theta \in J$ and if $M = \theta(R) + R$ generates mod-R, then $\theta(1) \in Q_c$.

Proof: If $\theta \in J$, then $M = \theta(1) \in E = E(R)$. Since $M = xR + R$ generates mod-R, then by Proposition 3.1A there exists $\lambda_1, \lambda \in \Lambda$ such that

$$\lambda_1(x) + \lambda(1) = 1 \tag{3.1}$$

$$\lambda_1(M) + \lambda(M) = R \tag{3.2}$$

Now (3.1) can be rewritten:

$$\lambda_1\theta(1) + \lambda(1) = 1 \tag{3.3}$$

which shows that $\ker \theta \cap \ker \lambda = 0$. Using Chap. 2, property (F), $\theta \in J \Rightarrow \ker \lambda = 0$.

Since $H \approx \lambda(M)$ generates mod-R, then by Theorem 2.2, $S = \text{End}_R H \approx R$ canonically (see Remark 2.3), and therefore $R \approx \text{End}_R M$ canonically. Since $\lambda(M) \subseteq R \subseteq M$, it follows that there exists $r \in R$ such that $\lambda(m) = mr$ for all $m \in M$, and in fact necessarily $r = \lambda(1)$. In particular $\lambda(x) = rx \in R$. Since $\ker \lambda = 0$, r is a regular element of R, so that $x = \theta(1) = \lambda(x)r^{-1} \in Q_c$ as asserted, for all $\theta \in J$.

The next result shows that the (upper) FPF hypothesis stunts the growth of the singular submodules of between modules of E over R.

3.3 COROLLARY If R is essentially upper FPF, then

$$\text{sing } E = \text{sing } Q_c = JE \subseteq Q_c$$

In particular, $JQ_c \subseteq JQ \subseteq Q_c \subseteq Q$.

Proof: Since $E = \Lambda 1 = \{\lambda(1) \mid \lambda \in \Lambda\}$, if $x \in E$, then $x = \lambda(1)$ for some λ, and then $x \in \text{sing } E$ iff $\ker \lambda \subset_{ess} E$, that is, iff $\lambda \in J$, by Utumi's theorem (F). Then $x \in Q_c$ by Proposition 3.2.

Henceforth S will denote sing E. Each $q \in Q$ defines $q_d \in \Lambda$ by multiplication, and we set $Q_d = \{q_d \in \Lambda \mid q \in Q\}$. Thus, $Q \hookrightarrow \Lambda$ under

$q \longrightarrow q_d$. Furthermore, since $S \subseteq Q_c \subseteq Q$, then by Corollary 3.3,

$$S_d = \{s_d \in \Lambda \mid s \in S\} \subseteq J$$

Since $Q = \text{End}_{\Lambda}E$, it is clear that Q_d = center Λ and $Q_d \cap \Omega = 0$.

3.4 COROLLARY Under the same assumption, Ω is an ideal of J, $J = \Omega + S_d$, $\Omega \cap S_d = 0$, and $J/\Omega \approx S_d$.

Proof: Now $\lambda \in \Lambda \Rightarrow \lambda(1) = q \in Q_{c\ell}(R)$ by Proposition 3.2, so $\omega = \lambda - q_d$ satisfies $\omega(1) = 0$, so $\omega \in \Omega$. Since $\lambda \in J$, then $(\ker \lambda) \cap Q = q^1 \subseteq_{ess} Q$, so $q = \lambda(1) \in S$ as stated. Thus, $J = \Omega + S_d$. Since $S_d \subseteq Q_d$ = center Λ, then Ω is an ideal of J. Since $\Omega \cap Q_d = 0$, the rest is clear.

Regarding the situation of Corollary 3.3, we have $S = \text{sing } Q_c = \text{sing } E = \text{sing } Q_2$ for any between ring Q_2 of Q and Q_c. The next lemma axiomatizes this situation. [The proof utilizes the discussion preceding Proposition 2.9, e.g., that $Q = Q_{max}(Q_2)$, that E is the injective hull of Q_2 in mod-Q_2, and that sing Q_2 (or sing E) is the same set regarding Q_2 (or E) as an R- or as a Q_2-module.]

3.5 LEMMA Let Q_2 be a between ring of Q and Q_c such that $S = \text{sing } E = \text{sing } Q_2$, let M be a between module of E over Q_2, and set $H = \{q \in Q \mid qM \subseteq Q_2\}$. If $M^* = \text{Hom}_{Q_2}(M,Q_2)$ is faithful, then H is faithful.

Proof: Since E is the injective hull of Q_2, and since $\Lambda = \text{End}_{Q_2}E$, then every $f \in M^*$ is induced by $\lambda \in \Lambda$. Partly conversely, any $\gamma \in J$ induces $\gamma \in M^*$, inasmuch as $\gamma(M) \subseteq Q_2$. If $\bar\lambda \in M^*$, then $\bar\lambda$ is induced by some $\lambda \in \Lambda$, and since $\lambda(1) = q \in Q_2$, then $\lambda = \omega + q_d$, with $\omega \in \Omega$ and $q \in Q_2$. (See Corollary 3.4.) Since $\bar\omega, \bar\lambda \in M^*$, then for all $m \in M$,

$$xq = q_d(m) = (\bar\lambda - \bar\omega)(m) \in Q_2$$

so $q \in H$. Since Q_2 is essential in E, then H is essential in Q_2, and so $Ha = 0 \Rightarrow a \in \text{sing } Q_2$. Since ker a_d is essential, then $a_d \in J$; hence $aM \subseteq Q_2$, so $a \in H$. Now qa, so $\bar\lambda a_s = \bar\omega a_s$. But $ma \in Q_2$, so

$$\bar\omega a_s(m) = \omega(ma) = 0 \quad \forall m \in M$$

since $\omega \in \Omega$. Thus $\lambda a_s = 0$, that is, $\lambda a = 0$ for all $\lambda \in M^*$. Thus, M^* faithful $\Rightarrow a = 0$; hence H is faithful.

3.6 COROLLARY Under the assumptions of Lemma 3.5, if M generates mod-Q_2, then $M \subseteq Q$. Furthermore, then $H \approx M^*$ and $M \approx H^*$ are finitely generated projective over Q_2, and $Q_2 = \mathrm{End}_{Q_2} M = \mathrm{End}_{Q_2} H$ canonically.

Proof: By Corollary 3.1B, M^* is faithful; hence $H = \{q \in Q \mid qM \subseteq Q_2\}$ is faithful by Lemma 3.5. Moreover, by the proof of Lemma 3.5, every $\overline{\lambda} \in M^*$ is induced by some $\lambda = \omega_\lambda + q_d \in \Lambda$ with $\omega_\lambda \in \Omega$ and $q \in H$, so the trace ideal of M in Q_2 is $T = \Sigma_{\overline{\lambda} \in M^*} \lambda(M) = \Sigma\, \omega_\lambda(M) + \Sigma_{q \in H}\, qM \subseteq S + HM$. Since M generates mod-Q_2, then $T = Q_2$, hence $Q_2 = S + H(M)$. But $S \subseteq H \subseteq HQ_2 \subseteq HM$, hence $HM = Q_2$. It is intuitively evident that this implies that H generates mod-Q_2, and we provide the details for later use. Let $\Delta = \{\delta \in \Lambda \mid \delta(1) \in M\}$. (Thus $M = \Delta Q_2$.) Then

$$Q_2 = \Delta Q_2 H = \Delta H = \sum_{\substack{\delta \in \Delta, h \in H \\ \text{(finite sums)}}} \delta(h)$$

Since $\delta(1) \in M$, then $\delta(h) \in Q_2$, so that each $\delta \in \Delta$ induces $\overline{\delta} \in H^*$, and hence Q_2 is the trace ideal of H in Q_2; that is, H generates mod-Q_2.

By Proposition 3.1A, Q induces H^*. If $m = \lambda(1) \in M$, then $mh = \lambda(h) \in Q$ for all $h \in H$, so λ induces an element $\overline{\lambda}$ of H^*, and by the proof of Lemma 3.5, since Q induces H^*, then there exists $q \in Q$ such that $\lambda(h) = qh$ for all $h \in H$. Since H is dense in Q, then $\lambda = \omega + q_d$ with $\omega \in \Omega$, i.e., $m = \lambda(1) = q \in Q$, so $M \subseteq Q$.

This implies that M is also a rank-1 generator of mod-Q_2, hence by Theorem 2.2, M and H are finitely generated projective, and

$$Q_2 \approx \mathrm{End}_{Q_2} M \approx \mathrm{End}_{Q_2} H$$

Finally, since $M \subseteq Q$ and Q induces M^*, then $H \approx M^*$ canonically, and then finite projectivity of M implies that $M \approx H^*$ canonically.

3.7 PROPOSITION If R is essentially upper FPF, then Q is injective (both as an R and as a Q module).

Proof: By Corollary 3.3, sing E = sing $Q_c \subseteq Q$, so the hypothesis of Lemma 3.5 holds for $Q_2 = Q$. Furthermore, by Proposition 2.9, Q is essentially upper FPF; hence if $\lambda \in \Lambda$, then $M = \lambda Q + Q$ generates mod-Q. Thus, $M \subseteq Q$ by Corollary 3.6, and therefore $\lambda Q \subseteq Q$ for all $\lambda \in \Lambda$. Since $Q \subseteq E =$

$\Lambda 1 = \Lambda R = \Lambda Q \subseteq Q$, then $Q = E$ is injective (in both mod-R and mod-Q) by Property (B) of Chap. 2.

3.8 PROPOSITION If R is pre-FPF, then R is classically upper FPF.

Proof: If M is a finitely generated between module of Q_c and R, then $H = cM \subseteq R$ for some regular $c \in R$. Since H is finitely generated faithful, then H, whence M, generates mod-R.

3.9 COROLLARY If R is pre-FPF, then R is integrally closed in Q_c.

Proof: Apply Proposition 3.8 to Corollary 2.8.

If I is a right ideal, then I is *a complement right ideal* iff I is maximal with respect to $I \cap K = 0$ for some right ideal K; equivalently, I is *(essentially) closed* in R in the sense that I has no proper essential extension in R. For if I is a complement to K, then any essential extension I' of I has $I' \cap K = 0$. Conversely, if I is closed and if K is a complement of I, then by Zorn's lemma, there is a complement I_1 of I containing I, and it follows easily that $I_1 \underset{ess}{\supset} I$. Thus, $I_1 = I$ is a complement.

3.10 COROLLARY If R is pre-FPF and quotient-injective, then every complement ideal of R is generated by an idempotent.

Proof: Let I be any ideal of R, and let E(I) be an injective hull taken in Q. Since $Q = E(I) \oplus X$ for some X in mod-R, then there exists an idempotent element $e = e^2 \in Q = \text{End}\,_R Q$ with $eQ = E(I)$, $(1 - e)Q = X$. By Corollary 2.8, $e \in R$, hence $eR = eQ \cap R$ is the maximal essential extension of I in R. If I is a complement ideal, then I is closed in R, hence $I = eR$.

Rings with the property stated in the corollary are studied in Refs. 12 and 13, where they are called CS (complements are direct summands) rings.

4. PROOF OF THEOREM B

In this section, let $\overline{Q} = Q/J$. Since $J \underset{\neq}{\supseteq}' \text{sing}\, Q \subseteq Q_c$ for any essentially upper FPF ring R by Corollary 3.3 and Proposition 3.7, we can form $\overline{Q}_c = Q_c/J$. Actually Corollary 3.3 shows that R upper-quotient FPF suffices for this. Clearly Q_c *is injective if and only if* $Q_c = Q$; for by Proposition 3.7 any essentially upper FPF ring R has injective Q. *Moreover,* $Q_c = Q$ *holds iff* $\overline{Q}_c = \overline{Q}$; and the latter is established in this chapter in six steps.

STEP 1 Every principal ideal $\overline{pQ}_c$ is projective and the annihilator $r_{\overline{Q}_c}(\overline{p})$ is generated by an idempotent, for any $\overline{p} \in \overline{Q}_c$.

Proof: $\overline{Q}$ is regular, hence unit regular since commutative, so every $\overline{p} \in \overline{Q}$ satisfies $\overline{p}^2\overline{u} = \overline{p}$ for a unit $\overline{u} \in \overline{Q}$. Then, $\overline{e} = \overline{pu}$ is an idempotent of $\overline{Q}$ which generates $\overline{pQ}$. Since Q is injective, idempotents of $\overline{Q}$ lift, so $pu = e + \gamma$ for $\gamma \in J$, and $e = e^2 \in Q$. By Corollary 2.8, $e \in Q_c$ and hence $\overline{e} \in \overline{Q}_c$, so $\overline{pQ}_c \approx \overline{eQ}_c$ (proving that $\overline{pQ}_c$ is projective), and clearly $r_{\overline{Q}_c}\overline{e} = (1 - \overline{e})\overline{Q}_c = r_{\overline{Q}_c}(\overline{pu}) = r_{\overline{Q}_c}(\overline{p})$.

STEP 2 $Q_{c\ell}(\overline{Q}_c) = \overline{Q}_c$.

Proof: If $\overline{p}$ is regular in $\overline{Q}_c$, then in the notation above, $\overline{e} = 1$; hence $\overline{p}$ is a unit of $\overline{Q}$, so p is a unit of Q, and therefore p is regular in Q_c. Since $Q_{c\ell}(Q_c) = Q_c$, then $p^{-1} \in Q_c$ and $\overline{p}^{-1} \in \overline{Q}_c$.

STEP 3 $\overline{Q}_c$ is a regular ring.

Proof: Show that any commutative ring having the property stated in Step 1 has regular $Q_{c\ell}$, and apply Step 2. (Cf. lemma 3E of Ref. 1.)

STEP 4 $\overline{Q} = Q_{max}(\overline{Q}_c)$.

Proof: First, we show that $\overline{Q} \subseteq Q_{max}(\overline{Q}_c)$, and for that it suffices to show that $\overline{Q}_c$ is dense in Q as a $\overline{Q}_c$-module. Since $\overline{Q}_c$ is regular, it suffices to prove that $\overline{Q}$ is an essential extension of $\overline{Q}_c$. Suppose $\overline{xQ}_c \cap \overline{Q}_c = 0$ for $\overline{x} \in \overline{Q}_c$. Then $xQ_c \cap Q_c \subseteq J$, since if $xq_1 = q_2 \in Q_c$ and $q_1 \in Q_c$, then $\overline{q}_2 = 0$, so $q_2 \in J$.

Let $M = xQ_c + Q_c$ and $H = M^* \subseteq Q_c$ as in Corollary 3.6, so that MH = Q. Now $xH = xQ_c \cap Q_c \subseteq J$, so

$$Q = MH = xH + H \subseteq J + H \subseteq H$$

and H = Q implies that $x \in Q_c$, so $\overline{xQ}_c \cap \overline{Q}_c = 0 \Rightarrow \overline{x} = 0$. This proves

$$\overline{Q}_c \subseteq \overline{Q} \subseteq Q_{max}(\overline{Q}_c)$$

By Proposition 2.9, this implies that $Q_{max}(\overline{Q}) = Q_{max}(\overline{Q}_c)$. Since $\overline{Q}$ is injective, then $\overline{Q} = Q_{max}(\overline{Q}_c)$ as stated.

STEP 5 $\overline{Q}_c$ is quotient-FPF.

Proof: Let $\overline{M}$ be a finitely generated between module for $\overline{Q}$ over $\overline{Q}_c$. If $\overline{x}_1, \ldots, \overline{x}_n$ generate $\overline{M}$. then $M = \Sigma_{i=1}^{n} x_i Q_c + J$ is the inverse image of $\overline{M}$ under $Q \longrightarrow \overline{Q}$. Since $\overline{M} \supseteq \overline{Q}_e$, then $M \supseteq Q_c + J = Q_c$ so $M = \Sigma_{i=1}^{n} x_i Q_c + Q_c$ is finitely generated over Q_c, hence generates mod-Q_c. Let H be as stated in Corollary 3.6, so that $MH = Q_c$ and $H = M^* \supseteq J$. Then $\overline{MH} = \overline{Q}_c$ and hence $\overline{M}$ generates mod-$\overline{Q}_c$.

FINAL STEP $\overline{Q} = \overline{Q}_c$.

Proof: By a proposition of Ref. 1, stated as Theorem 2.6A in Chap. 2, $Q_{c\ell}(\overline{Q}_c)$ is injective provided that $\overline{Q}_c \hookrightarrow Q_{max}(\overline{Q}_c)$ is a flat-epic. Since every finitely generated between module of Q over Q_c generates mod-$\overline{Q}_c$, hence it is finitely generated projective. Therefore $\overline{Q}_c \hookrightarrow \overline{Q}$ is a flat-epic by the sufficient condition stated in theorem 1E of Ref. 1, namely: every $b \in \overline{Q}$ belongs to a $\overline{Q}_c$-module M_b which generates mod-$\overline{Q}_c$. Since $Q_c(\overline{Q}_c)$ is injective, it coincides with $\overline{Q}_c = \overline{Q}$ by Steps 2 and 4.

This completes the proofs of Theorems A, B, and C.

5. QUOTIENT-INJECTIVE PRE-FPF RINGS ARE FPF

In this chapter we prove the theorem indicated by the title.

5.1 THEOREM The following are equivalent:

(1) R is FPF.

(2) R is quotient-injective and pre-FPF.

(3) R is essentially upper FPF and pre-FPF.

(4) Every finitely generated faithful ideal of R is projective, and every finitely generated between module of E(R) over R is projective.

Proof: (1) ⇒ (3) ⇒ (2) by Theorems A and B. Moreover, (3) ⇔ (4) since rank 1 generators are finitely generated, projective, and faithful, whereas the converse (Azumaya's theorem) holds in general (that is, without assuming rank 1).

(2) ⇒ (1) If M is finitely generated faithful over R, then $M \otimes_R Q$ is finitely generated faithful over Q since $R \hookrightarrow Q$, where $Q = Q_c = Q_m$. Let h be the canonical map $M \otimes_R R = M \longrightarrow M \otimes_R Q$, so K = ker h is the torsion submodule and $\overline{M} = M/K$ is torsion free. Then $\overline{M}$ embeds in $M \otimes_R Q$ canonically, and $\overline{M} \otimes_R Q = M \otimes_R Q$. Then by Proposition 3.1A, if $m_i \in M$ (i.e., $m_i \otimes 1$), $i = 1, \ldots, n$, generate $M \otimes_R Q$, there exist $f_i \in \mathrm{Hom}_Q(M \otimes_R Q, Q)$, $i = 1, \ldots, n$, such that $\Sigma_{i=1}^n f_i(m_i) = 1$. Write $f_i(m_i) = r_i t^{-1}$, with $r_i \in R$, $i = 1, \ldots, n$, and regular $t \in R$. So $t = \Sigma_{i=1}^n f_i(m_i)t \in R$, hence $f_i t$ induces $f_i' \in \mathrm{Hom}_R(\overline{M},R) = \overline{M}^*$, $i = 1, \ldots, n$. Therefore

$$P = \sum_{i=1}^{n} f_i'(\overline{M})$$

is a finitely generated faithful R-submodule of Q, so $P \approx cP \subseteq R$ for some regular element $c \in R$, and then pre-FPF implies that P generates mod-R. But $\overline{M}$ generates P, that is, $\overline{M}^n \twoheadrightarrow P$, hence $\overline{M}$ generates mod-R, and since $\overline{M} \twoheadrightarrow M$, then M generates mod-R. This shows R is FPF.

Remarks: 1. The proof uses the trivial fact that R pre-FPF implies that R is classically upper FPF. The converse holds for noetherian R since every dense ideal I in noetherian R contains a regular element. (See Ref. 1 for references.) Thus I finitely generated faithful in R implies that $c^{-1}I \supseteq R$ for some regular $c \in R$, hence $c^{-1}I$, whence I, generates mod-R.

2. If R is self-injective, then R has no finitely generated faithful ideals $\neq R$. For then R induces I^* for any ideal I, and I faithful implies $R = I^*$ canonically. Since R is FPF, I is projective and $I^*I = R$, so $RI = I = R$.

3. R is trivially pre-FPF when R has no finitely generated faithful ideals $\neq R$. In this case necessarily $R = Q_c$, since every regular element is a unit.

5.2 COROLLARY The following are equivalent:

(1) R is semihereditary and quotient-injective.

(2) R is reduced pre-FPF and quotient-injective.

(3) R is pre-FPF and Q_c is regular and self-injective.

(4) R is reduced FPF.

(5) R is semihereditary and $R \hookrightarrow Q_{max}(R)$ is a flat-epic.

(6) R is semihereditary and every finitely generated nonsingular R-module is projective.

Proof: (1) ⇔ (4) ⇔ (5) follow by Ref. 1, proposition 3B and theorem 4, and then (1) ⇔ (2) by Theorem 5.1. Also, (2) ⇔ (3) is trivial inasmuch as Q_c is injective over R iff self-injective. (Moreover, R reduced implies that $Q = Q_{max}(R)$ is regular, so Q_c injective implies that Q_c is regular. Conversely, Q_c regular implies that R is reduced.)

Cateforis proved for a semihereditary ring R that $R \hookrightarrow Q_m$ is a flat-epic iff every finitely generated nonsingular module is projective. [Goodearl generalized this by replacing "semihereditary" by nonsingular and "projective" by "embeddable in a projective" (see Ref. 1 for references).] This establishes (5) ⇒ (6).

5.3 COROLLARY If R is reduced FPF, then every finitely generated module M is a direct sum, $M = M_1 \oplus M_2$, where M_1 = sing M is the singular submodule and $M_2 = I_1 \oplus \cdots \oplus I_n$ is a finite direct sum of ideals of R.

Proof: M/M_1 is nonsingular and therefore projective, so $M_1 \hookrightarrow M$ splits. Since projectives over semihereditary rings have the required description by a theorem of Kaplansky (see Ref. 14, chapter 1), the proof is complete.

5.4 LEMMA Let $Q = Q_{c\ell}(R)$, J = rad Q, $P = J \cap R$, $\overline{R} = R/P$, and $\overline{Q} = Q/J$. Then $\overline{R} \hookrightarrow \overline{Q} \hookrightarrow Q_{c\ell}(\overline{R})$ canonically. If $\overline{Q}$ is injective, then $\overline{Q} = Q_{c\ell}(\overline{R})$.

Proof: $\overline{R} \hookrightarrow \overline{Q}$ since $J \cap R = P$. Also $\overline{Q} \subseteq Q_{c\ell}(\overline{R})$, since if $q \in Q$, $q = ab^{-1}$ with $a,b \in R \Rightarrow \overline{q} = \overline{a}\overline{b}^{-1}$ with $\overline{a},\overline{b} \in \overline{R}$, $\overline{b}$ regular. If $\overline{r} \in \overline{R}$ is regular in $\overline{R}$, then $\overline{r}$ is regular in $\overline{Q}$, since $\overline{rq} = 0$ for $q = ab^{-1} \in Q$ with $a,b \in R$ and b regular implies that $\overline{a} = 0$, so $\overline{q} = 0$. Thus, if $\overline{Q}$ is injective, then $\overline{r}^{-1}$ exists in $\overline{Q}$, so $\overline{Q} = Q_{c\ell}(\overline{R})$.

5.5 THEOREM If R is FPF, and P is the contraction of J = rad Q, then $\overline{R} = R/P \approx (R + J)/J$ is FPF reduced, hence semihereditary, with $Q_{c\ell}(\overline{R}) = \overline{Q} = Q/J$.

Proof: By Theorem A, Q is injective, and hence $\overline{Q}$ is a regular self-injective ring by Property (G) in Chap. 2. If $\overline{I} = I/P$ is finitely generated faithful over $\overline{R}$, then by injectivity of $\overline{Q}$, it follows that $\overline{QI} = \overline{Q}$. (For any injective E and finitely generated left ideal K, $\ell_E r_R K \subseteq EK$. See Ref. 8, chap. 23, 23.21). This implies that $QI + J = Q$, hence $QI = Q$ since J is a superfluous submodule of Q (see Ref. 8, chap. 18, 18.3-4). Therefore, I is a faithful ideal of R. Write $I = I_1 + P$ for I_1 finitely generated.

Then $IQ = I_1Q + PQ = I_1Q + J = Q$, so $I_1Q = Q$, and I_1 is thereby faithful. Then FPF $\Rightarrow I_1$ generates mod-R. But $I_1^*I_1 = R \Rightarrow \overline{I}_1^*\overline{I}_1 = \overline{R}$, where $\overline{I}_1^* = (I_1^* + J)/J \subseteq Q$. This shows that I generates mod-Q. Since $\overline{Q} = Q_{c\ell}(\overline{R})$ by Lemma 5.4, then $\overline{R}$ is quotient-injective, and as we have just shown, pre-FPF, so R is FPF by Theorem 5.1, and semihereditary by Corollary 5.2.

5.6 THEOREM (Partial Converse) If R is quotient-injective and if $\overline{R} = R/P$ is FPF, then $R' = R + J$ is FPF.

Proof: Since $Q_{c\ell}(R') = Q_{c\ell}(R) = Q$ and $R'/J \approx \overline{R}$, it suffices to prove the theorem for the case $R \supseteq J$. Since Q is injective, by Theorem 5.1 it suffices to prove R is pre-FPF. But if M is a finitely generated faithful ideal of R, then injectivity of Q implies that $MQ = Q$ (as in the proof of Theorem 5.5); hence $\overline{QM} = \overline{Q}$, so $\overline{M}$ is finitely generated faithful over $\overline{R}$, hence $\overline{M}^*\overline{M} = \overline{R}$ with $\overline{M}^* \subseteq \overline{Q}$, by Lemma 5.4. If H is the inverse image in Q of $\overline{M}^*$, then $HM + J = R + J = R$. But $H \supseteq R \supseteq J$, so $HM \supseteq M \supseteq MJ$. Since $MQ = Q$ and $JQ = J$, then $MJ = J$, whence $HM = HM + J = R$. Thus, M generates mod-R, and R is therefore FPF.

A ring R is a *sandwich ring* if $R \supseteq J = \text{rad } Q_c$. Clearly, $R + J$ is always a sandwich ring.

5.7 COROLLARY If R is a quotient-injective sandwich then R is FPF iff $\overline{R} = R/\text{rad } Q$ is FPF.

Proof: If R is FPF, then so is $\overline{R}$ by Theorem 5.5. If R is a sandwich ring, then $R = R + J$, and then the converse holds by Theorem 5.6.

5.8 PROPOSITION If R is FPF and if P is nil, then $\overline{R}$ is a sandwich ring.

Proof: Since $J = PQ$, then J is nil, so $P = N(R) = N(Q) = J \subseteq R$.

5.9 PROPOSITION Consider the properties:

(1) R is a sandwich ring.

(2) $P = R \cap \text{rad } Q$ is an ideal of $Q = Q_c$.

(3) $P = tP$ for all regular $t \in R$.

(4) P is a waist of R.

Then (1) $\Leftrightarrow$ (2) $\Leftrightarrow$ (3). Moreover, if Q is local then (1) through (4) are equivalent conditions.

(A submodule P of M is a *waist* of M provided that every submodule of M not contained in P contains P.)

Proof: (1) $\Rightarrow$ P = rad Q is an ideal of Q. Then tP = P for all units t of Q, hence (2) $\Rightarrow$ (3). If $j \in J$ = rad Q, then $j = rt^{-1}$ with $r, t \in R$, and $jt = r \in P$, so (3) $\Rightarrow$ $j \in t^{-1}P = P$. Thus $J = P \subseteq R$, so (3) $\Rightarrow$ (1).

Next assume (4) holds. Now t regular $\Rightarrow t^{-1} \in Q \Rightarrow t \notin J \Rightarrow t \notin P$, so (4) $\Rightarrow P \subseteq tR$. Write $P = tP'$ for an ideal $P' \supseteq P$. Then Q local and J maximal in Q $\Rightarrow$ P prime $\Rightarrow$ P' = P, so (3) holds. Conversely, assuming (3), then $r \in R\backslash P$ satisfies $rP = P \subseteq rR$, so (4) holds.

5.10 THEOREM [1] A. If R is local, then R is FPF iff $Q = Q_c$ is injective and the zero divisors of R form a waist P of R such that R/P is a VR.

B. Any FPF local ring R is a sandwich ring: rad Q = P.

Proof: A is theorem 9B of Ref. 1, and then B follows from Proposition 5.9.

6. CFPF = FSI

We prove the theorem indicated by the title of this chapter. One way has been established by Theorem A proved in Chap. 4. For the converse, since any factor ring of an FSI ring is FSI, and therefore a finite product of rings of the same three types,* then every factor ring of an indecomposable FSI ring has this property. Moreover, a finite product of CFPF rings is CFPF, so in order to prove FSI $\Rightarrow$ CFPF it suffices to prove that every indecomposable FSI ring is FPF. Now AMVR's and Prufer rings are FPF, so it reduces just to showing FPF for a torch ring R. But over such a ring, every finitely generated faithful module M has a projective faithful direct summand [3], so M generates mod-R by Azumaya's theorem.

Moreover, Vámos has communicated to me the following result deducible from Ref. 3:

6.1 THEOREM (Vámos) If R is FSI, and if M is a faithful finitely generated R-module, then there exist ideals $I_1, \ldots, I_n$ of R and ideals $J_1, \ldots, J_n$, such that $J_i \supseteq I_i$ and J_i/I_i is projective in R/I_i with "one and one-half generators," i = 1, ..., n, and $M = \oplus_{k=1}^n J_k/I_k$.

The proof of this will appear in a subsequent publication by Vámos.

---

*Described above Theorem 1.6.

7. FPF RINGS WITH SEMILOCAL QUOTIENT RINGS

A ring A is *semilocal* if A/J is a semisimple artinian ring, where J = rad A. If A is commutative, this happens iff A/J is a finite product of fields. In this section, we characterize the structure of FPF rings with semilocal Q_c. By Proposition 7.4, these are finite products of uniform FPF rings. A uniform FPF ring R is characterized in Theorem 7.7 by the properties: (1) Q_c is injective and local; (2) every finitely generated ideal containing a regular element is projective.

R is *indecomposable* if $R = R_1 \times R_2$ implies $R_1 = R$ or $R_2 = R$; equivalently, R has no idempotents except 0 or 1.

As before, N(R) denotes the set of nilpotent elements of R, that is, the maximal nil ideal or *nil radical* or R. Then $\bar{R} = R/N(R)$ is reduced.

7.1 THEOREM If R is FPF then $N(R) = N(Q_c)$.

Proof: Corollary 2.8.

7.2 SPLITTING THEOREM ([1], theorem 10) An FPF ring R splits, $R = R_1 \times R_2$, where R_1 is semihereditary and $N(R_2)$ is essential in R_2. (Possibly R_1 or $R_2 = 0$.)

7.3 COROLLARY Any indecomposable FPF ring R is either a Prufer domain or else has essential nil radical.

Proof: If N(R) is not essential, then R is semihereditary, and R indecomposable implies that R is a Prufer domain.

The product theorem [2] implies that any product $\Pi_{i\in I} R_i$ of commutative rings R_i is FPF iff each R_i is FPF.

7.4 PROPOSITION R is FPF with semilocal Q_c iff $R = R_1 \times \cdots \times R_n$ is a finite product of indecomposable FPF rings R_i with local $Q_i = Q_{c\ell}(R_i)$, $i = 1, \ldots, n$. Moreover, either R_i is a Prufer domain, or else $N_i = N(R_i) = N(Q_i)$ is an essential ideal of R_i, hence of Q_i, $i = 1, \ldots, n$.

Proof: We use the fact that a finite product $R = \Pi_{i=1}^n R_i$ of rings is FPF iff R is FPF. This can be deduced from the product theorem [2], but we prove it directly. If M_i is finitely generated (resp. faithful) over R_i, then

$$M = M_i + \prod_{\substack{j=1 \\ j \neq i}}^{n} R_j$$

is a module over R, and M_i is finitely generated (resp. faithful) over R_i iff M has that property over R. Thus, if R is FPF, then M generates mod-R, and then for the trace ideals we have

$$\text{Trace}_R\ M = R \Rightarrow \text{Trace}_{R_i}\ M = R_i$$

so M_i generates mod-R_i, proving that R_i is FPF, i = 1, ..., n. Conversely, if M is *any* R-module, then $M = \Pi_{i=1}^{n} M_i$ with $M_i = Me_i$, where e_i is the identity of R_i, i = 1, ..., n. Clearly M is finitely generated faithful over R iff M_i is the same over R_i, i = 1, ..., n. Thus, R_i FPF, i = 1, ..., n $\Rightarrow$ M_i generates mod-R_i, so M generates mod-R, and therefore R is FPF.

Now suppose R is FPF with semilocal $Q = Q_c$. Since Q is injective by Theorem A, then idempotents of Q/J lift, where J = rad Q, and hence Q semilocal implies that $Q = \Pi_{i=1}^{n} Q_i$, where $Q_i = e_iQ$ is a local ring such that rad $Q_i = e_iJ$, i = 1, ..., n. By Corollary 2.8, we have $e_i \in R$, hence $R = \Pi_{i=1}^{n} R_i$, where $R_i = Q_i \cap R$. Since Q_i has no nontrivial idempotents, then R_i is indecomposable and Corollary 7.3 completes the proof.

7.5 COROLLARY If R is FPF with semilocal $Q = Q_c$, then $\overline{R} = R/P$ is a finite product of Prufer domains, where P = (rad Q) $\cap$ R.

Proof: Write $R = \Pi_{i=1}^{n} R_i$, as in Proposition 7.4. Then $P = \Pi_{i=1}^{n} P_i$, where $P_i = P \cap R_i$, i = 1, ..., n. Since R/P is FPF by Theorem 5.5, then R_i/P_i is an FPF domain, hence Prufer, i = 1, ..., n, so $\overline{R} = \Pi_{i=1}^{n} R_i/P_i$ is a finite product of Prufer domains.

R is quasi-Frobenius (QF) iff R satisfies the following equivalent conditions:

(QF1) R is injective and artinian.

(QF2) R is injective and noetherian.

(QF3) R is injective and has acc $\perp$.

Any QF ring is PF. [See Ref. 8, chapters 24-25 for references. Also, (QF3) follows from Ref. 15.] A theorem of Tachikawa [16] states that any (left) perfect (right) FPF ring is (right) PF. Any (right and left) PF (left) perfect ring is QF by a theorem of Osofsky [11].

We can now give a short proof of Endo's theorem (stated in Ref. 1 as theorem 11), without recourse to Robson's theorem [1]. The proof however uses Robson's idea.

7.6 ENDO'S THEOREM (Redux) If R noetherian and FPF, then $R = R_1 \times R_2$, where R_1 is a finite product of Dedekind domains and R_2 is QF.

Proof: Since $Q = Q_c$ is injective and noetherian, then Q is QF by (QF2). We reduce to the case Q is local by Theorem 7.3, and we may suppose $N \subset_{ess} R$ by the splitting theorem. Now N = rad Q is nilpotent, and $V = N/N^2$ is the unique simple Q-module (via Q being local QF). Since N/N^2 is a finitely generated R-module, then $Q/N \approx V$ is finitely generated over R, and similarly, so is N^i/N^{i+1} for every i (Robson's idea). Then Q is a finitely generated R-module, which forces Q = R, since $Q = Q_c$.

A module M is *uniform* iff the following equivalent conditions hold:

(U1) Any two submodules $\neq 0$ have intersection $\neq 0$.

(U2) E(M) is indecomposable.

It is known that any indecomposable injective E has local endomorphism ring $\Lambda = \text{End } E_R$. Also by Property (F) of Chap. 2, rad $\Lambda = \{\lambda \in \Lambda \mid \ker \lambda \neq 0\}$.

R is a *uniform ring* if R is a uniform R-module. In a uniform ring R, every finitely generated dense ideal contains a regular element. [See Theorem 7.7(3).]

7.7 THEOREM The following are equivalent:

(1) R is uniform FPF.

(2) R is FPF and Q_c is local.

(3) R is uniform, Q_c injective, and every finitely generated ideal I containing a regular element is projective (resp. generates mod-R).

Proof: For an R-module M, E(M) is indecomposable iff End E(M) is local iff M is uniform. Since FPF $\Rightarrow E(R) = Q_c$, (1) $\Rightarrow$ (2). Also, (1) $\Rightarrow$ (3) since FPF $\Rightarrow$ I generates mod-R. (See Theorem 2.2 in this connection.)

(3) $\Rightarrow$ (1) By Theorem 4.1, R is FPF provided Q_c is injective and R is pre-FPF. Thus, assuming (3), if I is finitely generated faithful we must have that I contains a regular element, since if $\{x_i\}_{i=1}^n$ are generators of I, then $I^\perp = \cap_{i=1}^n x_i^\perp$. However, by uniformity of R, $\cap_{i=1}^n x_n^\perp = 0$ iff for some i, $x_i^\perp = 0$. Thus there is a regular $x_i \in I$, so I generates mod-R, hence R is pre-FPF as required.

R *pre-PF* means that every faithful (i.e., dense) ideal of R generates mod-R.

7.8 COROLLARY Let R be uniform pre-PF. Then every dense (i.e., faithful) ideal I is finitely generated projective, and contains a regular element.

Proof: By pre-PF, I generates mod-R, so I is finitely generated projective by Theorem 2.2, and contains a regular element since R is uniform (Theorem 7.7).

7.9 COROLLARY (1) R is FPF with semilocal $Q = Q_c$ iff $R = \Pi_{i=1}^{n} R_i$ is a finite product of uniform FPF rings.

(2) When this is so, then the following conditions on a finitely generated ideal I are equivalent:

(a) I is dense.

(b) I generates mod-R.

(c) I is a faithful projective.

(d) I contains a regular element.

Proof: (1) follows from Proposition 7.4 and Theorem 7.7. Moreover, (2) also follows from Theorem 7.7, since any ideal $I = \Pi_{i=1}^{n} I_i$ for ideals I_i of R_i, $i = 1, \ldots, n$; moreover, I has one of the properties (a) through (d) iff I_i has the same property as an ideal of R_i, $i = 1, \ldots, n$.

Let zd(R) denote the set of all zero divisors of R.

7.10 THEOREM Let R be a uniform FPF ring. Then P = zd(R) is a prime ideal such that

$$R + J = \bigcap_{M \in S} R_M \tag{7.1}$$

where S is the set of prime ideals of R containing P, and

$$J = \text{rad } Q = PQ = PR_M \tag{7.2}$$

where $Q = Q_c$. Therefore, R is a sandwich ring iff

$$R = \bigcap_{M \in S} R_M \tag{7.3}$$

Proof: By Theorem 7.7, Q is a local ring, and Q\J consists of the units of Q. Since R is uniform, P is a prime ideal of R, so $Q = R_P$ has radical $J = PR_P = PQ$, and $J \cap R = P$.

As stated in the proof of Proposition 2.9, any between ring of Q over R is FPF, so R_M is FPF for all $M \in S$. Since $\overline{R}$ is Prufer by Theorem 7.5, then $\overline{R} = \cap \overline{R}_N$ where N ranges over all prime ideals of $\overline{R}$, hence

$$\overline{R} = \bigcap_{M \in S} \overline{R}_{\overline{M}} \qquad \overline{M} = M/P, \ \forall M \in S \tag{7.4}$$

By Theorem 5.10, the set $zd(R_M)$ of zero divisors of R_M is a waist J. Yet obviously $zd(R_M) = PR_M$, that is, $J = PR_M$, so (7.2) holds. This fact together with the canonical isomorphisms $R_M/PR_M \approx \overline{R}_{\overline{M}}$ proves that (7.1) holds in view of (7.4). [Specifically, (7.4) states that $\cap_{M \in S} R_M \subseteq R + J$, whereas the reverse inclusion comes from $J \subseteq R_M$.] For the last statement, we note that $J \subseteq R$ iff (7.3) holds.

Remarks: 1. If R is a subring of $Q = Q_c$ such that $P = J \cap R$ is maximal, then Q injective implies $\overline{Q} = Q_{c\ell}(\overline{R}) = \overline{R}$, so then $Q = R + J$. As noted elsewhere this bizarre arrangement cannot hold for $R \neq Q$ when, e.g., R is noetherian or local or indeed whenever $J = PQ = P$, for example when $J = N(Q)$. (For R is a sandwich ring in these cases.)

2. If P is a prime ideal of R, and if T is a subring of Q_c containing R_P maximal with respect to the property that $(\text{rad } T) \cap R_P = PR_P$, then T is a valuation ring when R is a domain. (This result appears somewhere in Bourbaki's commutative algebra.)

8. FPF RINGS WITH PF QUOTIENT RINGS

A ring A is right PF provided that one of the following equivalent conditions holds:

(PF1) Every faithful right A-module is a generator of mod-A.

(PF2) A is injective in mod-A and right Kasch.

(PF3) A is an injective cogenerator of mod-A.

(PF4) A is a semilocal right self-injective ring with essential right socle $S_r = \text{soc}_r A$.

(PF5) $S_r = V_1 \oplus \cdots \oplus V_n$ for finitely many simple modules (right ideals) and $A = e_1A \oplus \cdots \oplus e_nA$, where e_i is a primitive idempotent such that e_iA is the injective hull of V_i, $i = 1, \ldots, n$.

See Ref. 8, chapters 24-25 for references.

That e_i is primitive means that e_iAe_i is a local ring, and consequently,

e_iA/e_iJ is a simple module, $i = 1, \ldots, n$, where $J = \text{rad } A$. If A is PF, then in (PF5), $e_iA/e_iJ \approx V_{\pi(i)}$, where π is an element of the symmetric group on $1, \ldots, n$.

The equivalence of (PF1) through (PF5) is a theorem of G. Azumaya, B. Osofsky, and Y. Utumi.

8.1 THEOREM (1) R is PF iff $R = R_1 \times \cdots \times R_n$ for local self-injective PF rings R_i, $i = 1, \ldots, n$.

(2) Moreover, R is a local PF ring iff R is self-injective with essential simple socle.

Proof: (1) follows from Proposition 6.3 since R is PF $\Rightarrow R = Q_c$ is semilocal by (PF4). Also, (2) follows from (PF5).

If a ring A satisfies the ascending (descending) chain condition on annihilator right ideals, then we say A satisfies acc $\perp$ (resp. dcc $\perp$). The left-right symmetry is $\perp$ acc ($\perp$ dcc). Clearly $\perp$ dcc $\Leftrightarrow$ acc $\perp$, and for commutative R we have acc $\perp$ iff dcc $\perp$.

8.2 PROPOSITION If R is pre-PF, or pre-FPF with acc $\perp$, then Q_m is Kasch (i.e., has no faithful ideals $\neq Q_m$). If, in addition, Q_c is injective (e.g., when R is FPF), then Q_c is PF in the first instance, and QF in the second.

Proof: Proposition 1G of Ref. 1 implies the first part, and then Q_c injective $\Rightarrow Q_c = Q_m$ is PF by (PF2). Now acc $\perp$ in R $\Rightarrow$ acc $\perp$ in Q_c; hence Q_c is QF by (QF3), Chap. 7.

A ring R is a (right) Σ-Levitzki ring (or right Σ-ring for short) if E(R) is Σ-injective. This is equivalent to saying that R has the acc on the set $A_r(E(R),R)$ of (right) ideals of R which are annihilators of subsets of E(R) (see Ref. 15). We say R is a (right) Δ-ring if R has dcc on these (right) ideals. It is known that any Δ-ring is a Σ-ring. More generally, for any injective R-module E, dcc on $A_r(E,R) \Rightarrow$ acc on $E_r(E,R)$ (theorem of Teply-Miller [17]). Moreover, by a theorem of Beck [18], R is a $\Sigma(\Delta)$-ring iff Q_c is noetherian (artinian) (also see Ref. 19). The next theorem states that the converse holds for FPF rings.

8.3 COROLLARY Any FPF Σ-ring R is a Δ-ring.

Proof: Q_c is QF, hence artinian by Proposition 8.2.

8.4 THEOREM Let R be FPF and suppose that R or Q_c is pre-PF. Then $Q_c = Q_1 \times \cdots \times Q_n$ is a finite product of local PF rings Q_i, $i = 1, \ldots, n$, and $R = R_1 \times \cdots \times R_n$, where $R_i = Q_i \cap R$, is an indecomposable ring, and $Q_{c\ell}(R_i) = Q_i$, $i = 1, \ldots, n$.

Proof: Since $Q = Q_c$ is injective by Theorem A, then Q is PF by Proposition 8.2, so Corollary 7.3 and Theorem 8.1 apply. See Corollary 7.9 for the converse.

In Ref. 20, Camillo and Fuller studied rings over which every faithful right module is a finitely generated flat module over its endomorphism ring, and proved these rings coincided with the right PF rings. Moreover, their proof *mutatis mutandis* suffices to classify the right FPF rings as those over which the finitely generated faithful module has the stated property.

9. NOTE ON THE GENUS OF A MODULE AND GENERIC FAMILIES OF RINGS*

The *genus* g(R) of a ring is the infimum g of all integers γ such that for all finitely generated generators M of mod-R, M^{γ} has a unimodular element; equivalently, there is an epic $M^{\gamma} \longrightarrow R$. (If no such exists, then set g = $g = \infty$.) [This defines the right genus $g^r(R)$, which we denote by g(R) for simplicity.] A family $F = \{R_i\}_{i \in I}$ of rings is *generic of (with) bound* B if there exists a function $B : \mathbb{Z}^+ \longrightarrow \mathbb{Z}^+$ such that for all modules M, if $\nu(M) < \infty$ is the minimal number of elements in any set of generators of M, and if M generates mod-R, then there is an epic $M^{B(\nu(M))} \longrightarrow R$. The product theorem [2] states that any product of a generic family of rings of bound B is a ring which is generic of bound B (considering a ring as a family with one member). For example, a family of rings each of genus $\leqslant g$ is generic with bound $\leqslant g$, where g also denotes the constant function. Moreover, *any* family of commutative rings is generic of bound $1_{\mathbb{Z}^+}$. The 2 × 2 theorem of Ref. 1 states that if R is a commutative ring of genus 1, then for any faithful module M with $\nu(M) = 2$, the product M^2 has a unimodular element. Thus, by the product theorem, the 2 × 2 theorem holds for any product of such rings.

A corollary of the product theorem is that any product $R = \Pi_{i \in I} R_i$ of a generic family of right FPF rings is right FPF. (In particular, the

*This section is reprinted from Ref. 1, p. 197.

product of any family of commutative FPF rings is FPF.) This implies that any product of self-basic right FPF rings, in particular any product of self-basic right PF rings, is right FPF of genus 1.

Another corollary to the product theorem states that if $\{R_i\}_{i\in I}$ is any family of commutative rings, and if there exist integers $n > 0$ and $g > 0$ with the property that for all $i \in I$ every finitely generated R_i-module of free rank $\geqslant n + 1$ has genus $\leqslant g$, then their product R has the same property: Every finitely generated R-module of free rank $\geqslant n + 1$ has genus $\leqslant g$. The FPF theorem for commutative R is just the case $g = 1$ and $n = 0$.

10. FP^{2}F AND CFP^{2}F RINGS AND THE "BIG" GENUS

Let FP^{2}F denote the condition in a ring R that all finitely presented faithful modules generate mod-R, and let CFP^{2}F mean that all factor rings of R are FP^{2}F. FP^{2}F local rings have been characterized analogously to FPF local rings (see Ref. 1, especially theorem 25 and the note following). A local ring R is CFP^{2}F iff R is a valuation ring, that is, a chain ring. (This and the other theorems on FP^{2}F rings quoted are proved in sec. 4 of Ref. 1.) CFP^{2}F is a local property of commutative rings, that is, R is CFP^{2}F iff every local ring of R is a VR. While local FP^{2}F implies FP^{2}F, the converse is not true. However, a domain R is FP^{2}F iff R is locally FP^{2}F. Any commutative flat-ideal ring, e.g., any semihereditary ring R, is CFP^{2}F. (R is a flat-ideal ring iff R is locally a valuation domain.)

PROBLEMS

1. Characterize commutative FP^{2}F and CFP^{2}F rings analogously to our characterization in Chap. 1 of FPF and CFPF rings.
2. Characterize right FPF rings.
3. Characterize right *and* left FPF rings. Do they have injective Q_{max}'s? (Cf. Ref. 21.) And is Q_{max} a flat epic of R? (This holds in all examples of FPF rings.)
4. Is every right FPF ring right slim in the sense that no infinite direct sum of copies of the same nonzero module cam embed in R? All presently known right FPF rings are slim; and a right self-injective ring R is right FPF iff R is right slim. (This will appear in Ref. 22.) Actually,

commutative FPF rings are *very* slim in the sense that M^2 cannot embed in R for any $M \neq 0$ since injectivity of Q implies that End ${}_R M^2$ is induced by a subring of Q. But the former is noncommutative, while the latter must be commutative along with Q.

5. Let G(R) denote the "big" genus of R, that is, in the definition of g(R) relax the condition on the generators of mod-R to permit *all* generators instead of just the finitely generated ones. Obviously, $g(R) \leqslant G(R)$, but how may they differ? If R is right noetherian, then $g^r(R) = G^r(R)$ by a result of Wiegand proved in Ref. 2. If R is commutative, then $g(R) \geqslant G(Q) = g(Q)$. Now $g(R) \geqslant g(Q)$ holds for any flat-epic $R \hookrightarrow Q$, whereas when Q is injective, then for any generator M of mod-Q, there a finitely generated faithful Q-submodule M_0 of M. (Hint: If $f_i \in M^*$, $x_i \in M$, are such that $\Sigma_{i=1}^n f_i x_i = 1$, then $M_0 = \Sigma_{i=1}^n x_i Q$ is faithful.) Then, since $M_0^{g(Q)} = Q \oplus X$ in mod-Q, we get an embedding $Q \hookrightarrow M^{g(Q)}$, and hence injectivity of Q implies that $M^{g(Q)} = Q \oplus Y$ in mod-Q, so $G(Q) \leqslant g(Q)$ as asserted.

So far, there are no examples proving that the big genus is strictly bigger than the genus.

6. If R is a commutative ring, and if every ideal of R has commutative endomorphism ring, then is $Q_{max}(R)$ injective? (The converse holds, as we have noted in Problem 4. Also note that for any R, every faithful ideal has commutative endomorphism ring.) This problem is problem 1 appended at the end of Ref. 1. Also see the text of Ref. 1 and Refs. 21 and 22 for additional problems on FPF rings. In Ref. 21 there is a very weird example of an FPF ring which shows that right FPF need not imply left. What happens if R is a prime right noetherian right FPF ring? Does R have to be left FPF in this case?

7. If R is right FPF, and if G is a finite group of automorphisms of R, then is the Galois subring $A = R^G$ also right FPF? This appears unlikely to hold in general. However, for a commutative ring R, an affirmative result does appear in Ref. 23 for a group G which is *pre-Galois* in the sense that R is finitely generated projective over A, and G is an independent group of automorphisms (see Ref. 24). Actually, the right FPF property is inherited for much more general extensions R/S such that R is left faithfully flat over S, and also a right projective generator over S. Moreover, for right self-injectivity to be inherited by S, it suffices for R to be a left flat right generator over S (called a *flat split-extension* in Ref. 23).

8. Characterize when RG is right FPF for a group G. This has been done by K. Louden (unpublished) for right PF; namely RG is right PF iff R is right PF and $|G| < \infty$. (Wilkerson [25] did this assuming $|G|$ invertible.) For a general right FPF ring R, almost nothing is known about when RG is right FPF, even assuming $|G|^{-1} \in$ R. However, when R is commutative FPF, one can show that QG is right FPF, where $Q = Q_c(R)$. (Hint: if M is any finitely generated QG-module, then the fact that R embeds in End_{QG} M shows that M is finitely generated over its endomorphism ring, and that implies that QG embeds in a finite product M^n, that is, M is cofaithful over QG, by the standard argument. [See, for example, Ref. 8, p. 68, prop. 19.14B(b).] Then, injectivity of QG implies that QG splits in M^n, so M generates mod-QG.)

REFERENCES

1. C. Faith, Injective quotient rings of commutative rings, in *Module Theory*, Lecture Notes in Math., Vol. 700, (C. Faith and S. Weigand, eds.), Springer-Verlag, Berlin-Heidelberg-New York, 1979.
2. ______, The genus of a module and generic families of rings, in *Antwerp Ring Theory Conference, 1978*, Lecture Notes in Pure and Applied Math., Vol. 51, Marcel Dekker, New York and Basel, 1979.
3. P. Vámos, The decomposition of finitely generated modules and fractionally self-injective rings, *J. London Math. Sco. (2), 16* (1977), 209-220.
4. W. Brandal, *Commutative Rings Whose Finitely Generated Modules Decompose*, preprint, U. of Tenn., Knoxville, 1977.
5. P. Vámos, Sheaf-theoretical methods in the solution of Kaplansky's problem, in *Applications of Sheaves* (M. P. Fourman, C. J. Mulvey, and D. S. Scott, eds.), Lecture Notes in Mathematics, Vol. 753, Springer-Verlag, Berlin-Heidelberg-New York, 1979.
6. R. Wiegand and S. Wiegand, *Commutative Rings over Which Finitely Generated Modules Are Direct Sums of Cyclics*, preprint U. of Nebraska Math. Dept., Lincoln, Nebraska, 1977.
7. J. Lambek, *Rings and Modules*, Blaisdell, New York, 1966; corrected reprint, Chelsea, New York, 1976.
8. C. Faith, *Algebra II, Ring Theory*, Grundl. der Math. Wiss., Vol. 191, Springer-Verlag, Berlin-Heidelberg-New York, 1976.
9. Y. Utumi, On quotient rings, *Osaka Math. J. 8* (1956), 1-18.
10. ______, Self-injective rings, *J. Algebra 6* (1967), 56-64.
11. B. L. Osofsky, Endomorphism rings of quasi-injective modules, *Canad. J. Math. 20* (1968), 895-903.

12. A. W. Chatters and C. R. Hajarnavis, Rings in which every complement ideal is a direct summand, *Quart. J. Math., Oxford (2) 28* (1977), 61-80.

13. A. W. Chatters and S. Khuri, Endomorphism rings of modules over non-singular CS rings, *J. London Math. Soc. (2), 21* (1980), 434-444.

14. H. Cartan and S. Eilenberg, *Homological Algebra*, Princeton U. Press, Princeton, 1956.

15. C. Faith, Rings with ascending condition on annihilators, *Nagoya Math. J. 27* (1966), 179-191.

16. H. Tachikawa, A generalization of quasi-Frobenius rings, *Proc. Amer. Math. Soc. 20* (1969), 471-476.

17. M. Teply and R. Miller, The descending chain condition relative to a torsion theory, *Pac. J. Math. 83* (1979), 207-219.

18. I. Beck, Σ-injective modules, *J. Algebra 21* (1972), 232-249.

19. C. Faith, Injective modules over Levitzki rings, Part I of these lectures.

20. V. P. Camillo and K. R. Fuller, Rings whose faithful modules are flat over their endomorphism rings, *Archiv. der Math. 27* (1976), 522-25.

21. S. Page, FPF rings and some conjectures of C. Faith, *Bull. Can. Math. Soc.* (to appear).

22. C. Faith and S. Page, *Azumaya-Morita Theory: Faithful Modules and Generators of Mod-R* (to appear).

23. C. Faith, Subrings of self-injective and FPF rings, *Proceedings of the Hudson Symposium*, SUNY, Plattsburgh (to appear).

24. ______, Dedekind's theorem on the independence of automorphisms revisited, in *Proc. Yale Symposium Honoring Nathan Jacobson*, Contemporary Mathematics, Amer. Math. Soc., Providence, 1982.

25. R. W. Wilkerson, Finite dimensional group rings, *Proc. of Amer. Math. Soc. 41* (1973), 10-16.

Supplementary Reading

S. Endo, Completely faithful modules and quasi-Frobenius algebras, *J. Math. Soc. Japan 19* (1967), 437-456.

R. E. Johnson and E. T. Wong, Quasi-injective modules and irreducible rings, *J. London Math. Soc. 36* (1961), 260-8.

S. Page, Regular FPF rings, *Pac. J. Math. 79* (1978), 169-175.

ABBREVIATIONS

PF = pseudo-Frobenius

FPF = finitely PF; FP^2F = finitely presented PF

CFPF = completely FPF; CFP^2F: all factor rings are FP^2F

FSI = fractionally self-injective

mod-R = the category of all (right) R-modules

QF = quasi-Frobenius

$Q_{c\ell}$ = the full, or classical, quotient ring

Q_{max} = the maximal quotient ring

rad-R = the Jacobson radical of R

sing R = the singular ideal of R

σ-cyclic = direct sum of cyclic modules; also called FGC(finitely generated modules are direct sums of cyclic modules)

σ-cyclic ring = one over which every finitely generated module is σ-cyclic

VR = valuation ring*

AMVR = almost maximal valuation ring

*A ring R such that for all $x,y \in R$, either $(x) \supseteq (y)$ or $(y) \subseteq (x)$. Equivalently, the ideals of R are linearly ordered. Chain ring is a variant name.

INDEX